Praise for
The Community Food Forest Handbook

"The increased interest in botanical sanctuaries, forest gardens, and nurtured edible gathering places is a sign that many are seeking to rekindle our relationship with these spaces as caretakers in these modern times. Learning how to nurture food forests builds community resilience, engagement, health, and stewardship. *The Community Food Forest Handbook* is perfect for townships, urban planners, landscape designers, community organizers, land trusts, permaculture enthusiasts, and foragers who aspire to dig in and seed our future. Catherine Bukowski and John Munsell have created a timely, well-researched guide that provides plenty of hands-on tools for advocacy and implementation based on diverse case studies from across the country. In the spirit of Robert Hart's classic *Forest Gardening*, it gives hope to see the community food forest trend rapidly resurging."

—**Susan Leopold**, executive director of United Plant Savers

"*The Community Food Forest Handbook* opens the door to a new, rapidly expanding approach to agroforestry in urban areas and communities. Rather than focus on the technical aspects of planting and production, Catherine Bukowski and John Munsell address sociological challenges inherent in planning and sustaining community food forests, as well as potential solutions. The result is a comprehensive resource for adapting practices traditionally applied to privately owned rural land for the enrichment of community-managed greenspaces."

—**Susan Stein**, director of
National Agroforestry Center of USDA Forest Service

"Wedding community renewal with agroecology, Bukowski and Munsell offer us a remarkably rich harvest of wisdom from a quarter of a century of insight and struggle in the community food forest movement. First fruits are everywhere consumed to nourish the spirit, reward unlikely heroes, and propitiate success. Partake of *The Community Food Forest Handbook*, and celebrate permaculture taking hold of America's imagination: from Seattle to Asheville, Syracuse to San Francisco, and in dozens of cities across the fruited plain, perennial culture is rising."

—**Peter Bane**, president of
Permaculture Institute of North America

"As communities seek both to grow food and to solve social and environmental problems, they need new insights into the ways in which people self-organize to initiate projects and sustain them in the long term. In *The Community Food Forest Handbook*, the authors offer a highly useful guide based on the collective wisdom of people and communities who are defining this practice as they develop it on the ground. The thoughtful analysis of planning strategies and numerous case studies of active projects help us all understand what community food forests are and can be for the future."

—**Steve Gabriel**, author of *Silvopasture* and
coauthor of *Farming the Woods*

"It's great to see a book about food forests / forest gardens which concentrates on community-scale projects. These need design and management not only for the growing system itself, but also for the human community that nurtures the forest and is often much neglected. *The Community Food Forest Handbook* does an excellent job of tackling the social issues and includes some highly informative case studies of community projects."

—**Martin Crawford**, director of Agroforestry Research Trust; author of *Trees for Gardens, Orchards, and Permaculture*

The Community Food Forest Handbook

The Community Food Forest HANDBOOK

How to Plan, Organize, and Nurture Edible Gathering Places

**Catherine Bukowski
and John Munsell**

Foreword by **LaManda Joy**

Chelsea Green Publishing
White River Junction, Vermont | London, UK

Copyright © 2018 by Catherine Bukowski and John Munsell
All rights reserved.

Unless otherwise noted, all photographs
copyright © 2018 by Catherine Bukowski

Photographs on pages 154, 187, and 196
courtesy of Jonathan H. Lee, www.subtledream.com.

No part of this book may be transmitted or
reproduced in any form by any means without
permission in writing from the publisher.

Project Manager: Alexander Bullett
Acquisitions Editor: Makenna Goodman
Developmental Editor: Fern Marshall Bradley
Copy Editor: Nancy Bailey
Proofreader: Rachel Shields Ebersole
Indexer: Shana Milkie
Designer: Melissa Jacobson
Page Composition: Abrah Griggs

Printed in Canada.
First printing July, 2018.
10 9 8 7 6 5 4 3 2 1   18 19 20 21 22

Chelsea Green Publishing is committed to preserving ancient forests and natural resources. We elected to print this title on paper containing 100% postconsumer recycled paper, processed chlorine-free. As a result, for this printing, we have saved:

70 Trees (40' tall and 6-8" diameter)
32,660 Gallons of Wastewater
32 million BTUs Total Energy
2,186 Pounds of Solid Waste
6,022 Pounds of Greenhouse Gases

Chelsea Green Publishing made this paper choice because we are a member of the Green Press Initiative, a nonprofit program dedicated to supporting authors, publishers, and suppliers in their efforts to reduce their use of fiber obtained from endangered forests. For more information, visit www.greenpressinitiative.org.

Environmental impact estimates were made using the Environmental Defense Paper Calculator. For more information visit: www.papercalculator.org.

Our Commitment to Green Publishing
Chelsea Green sees publishing as a tool for cultural change and ecological stewardship. We strive to align our book manufacturing practices with our editorial mission and to reduce the impact of our business enterprise in the environment. We print our books and catalogs on chlorine-free recycled paper, using vegetable-based inks whenever possible. This book may cost slightly more because it was printed on paper that contains recycled fiber, and we hope you'll agree that it's worth it. Chelsea Green is a member of the Green Press Initiative (www.greenpressinitiative.org), a nonprofit coalition of publishers, manufacturers, and authors working to protect the world's endangered forests and conserve natural resources. *The Community Food Forest Handbook* was printed on paper supplied by Marquis that contains 100% postconsumer recycled fiber.

Library of Congress Cataloging-in-Publication Data
Names: Bukowski, Catherine, author. | Munsell, John, author.
Title: The community food forest handbook : how to plan, organize, and nurture edible gathering places /
    Catherine Bukowski and John Munsell.
Description: White River Junction, Vermont : Chelsea Green Publishing, [2018]
    | Includes bibliographical references and index.
Identifiers: LCCN 2018007439| ISBN 9781603586443 (pbk.) | ISBN 9781603586450 (ebook)
Subjects: LCSH: Edible forest gardens. | Community gardens.
Classification: LCC SB454.3.E35 B85 2018 | DDC 635—dc23
LC record available at https://lccn.loc.gov/2018007439

Chelsea Green Publishing
85 North Main Street, Suite 120
White River Junction, VT 05001
(802) 295-6300
www.chelseagreen.com

To my parents, Sandra and Fran,
and my grandpa, Henry, for teaching me to love gardening
and understand the value of sharing good food.
To my husband, Jon, for accompanying me on the journey
and to all my family for their endless support.

—CJB

To E, O, and C. No way, no how, without you.

—JFM

Contents

Foreword   ix

Introduction   1

PART 1
Understanding Community Food Forests

1. Community Food Forests on the Rise   17
2. Systems Thinking for Community Food Forests   25
3. Capital Investments in Community Assets   35

PART 2
Meaningful Planning

4. Planning Fundamentals   53
5. Planning to Create Change   71
6. Rooting in History   81
7. The Dr. George Washington Carver Edible Park   95

PART 3
Functional Design

8. The Role of Agroecology   111
9. Allies in Creating and Managing Public Space   121
10. The Basalt Food Park   143

PART 4
Purposeful Community

11. Reflecting on Community   153
12. Building Social Systems   169

13 The Beacon Food Forest	183
14 Collaborative Leadership	197
15 The Bloomington Community Orchard	205
Conclusion: Looking Back, Moving Forward	**217**
Acknowledgments	*229*
Appendix: Goals, Visions, and Mission Statements Associated with Community Food Forests	*231*
Notes	*235*
Index	*241*

Foreword

The end of the Second World War marked the conclusion of an era in the United States when individuals and entire cities had a much more intimate and informed relationship with their food and where it came from. Pre-war food production was smaller, more decentralized, and closer to home. Many families, urban and otherwise, raised a few chickens and tended a home garden. A large percentage of the produce consumed in urban areas was grown within or close to city limits. In my adopted hometown of Chicago, there were neighborhoods known as "Celeryville" or "Pickletown" where businesses grew and processed food crops for the expanding urban, overwhelmingly immigrant, population. These farms provided food, yes, but also jobs, community, and a sense of belonging in a new land.

The juggernaut of the Second World War created efficiencies and businesses that had never existed pre-war, and, after the conflict ended, many industries retrofitted themselves to the challenge of a new American dream that didn't involve the day-to-day "drudgery" of food production. As the task of putting food on the table shifted from growing it yourself or buying it from a local farmer to weekly shopping trips to newly ubiquitous grocery stores, those precious home and locally based food production skills and networks began to fade away.

The last big hurrah of communal urban food production—the Victory Garden movement of the Second World War—was quickly deemed unnecessary as the food supply normalized in the decade immediately post-war. Citizens began the business of building a new society that, in retrospect, was founded on consumerism with a big hallmark being ready-made, cheap, industrial food.

Now the (dining) tables are turning. Collectively, we're rethinking the food system we thought we wanted and trying to relearn ways of connecting with food that once seemed mundane and outdated. We're reviving memories of inner-city greenhouses filled with trellised cucumbers, and hundreds of thousands of people fighting food shortages by tending community garden plots. These echoes of what has been and what can be are inspiring a new generation—I call it the "Greenest Generation"—to rethink our day-to-day relationship with how we obtain our food and the impact of that food on our planet.

This reimagining of our food system is why books like *The Community Food Forest Handbook* are so important. Through this people-centered handbook, Catherine Bukowski and John Munsell have given us the tools to create change in our communities and the inspiration to know that change is possible.

In Chicago, my own organization, Peterson Garden Project (PGP) has been able to harness the nostalgia of the Second World War Victory Garden by employing tactics from the original

FOREWORD

movement. Using empty urban land on a short-term basis—Pop-up Victory Gardens—we teach people how to grow their own food. In a borrowed culinary center, we then teach people how to cook it, too. A mission to "recruit, educate and inspire everyone to grow and cook their own food" has produced powerful results since 2010. By harnessing the energy of community, a very small team is able to facilitate the growth of 35,000 pounds of fresh, organic food annually. Almost 10 percent of this is grown specifically for food and nutrition partners in the neighborhoods where our many gardens are located. All of this is made possible by an education-first approach that trains gardeners and garden leaders to train others.

PGP has utilized an ABCD (Asset Based Community Development) model and worked with what we have—free land, eager new gardeners, and passionate neighbors—to create something special. Our legacy lives on in the education people can utilize and share for a lifetime. Organizations with a more permanent access to land have an even greater opportunity for longer-term food access and education by utilizing a food forest approach. And, lucky for them, *The Community Food Forest Handbook* exists.

This book isn't about just imagining a better future via food. It's practical, too. The sections on planning and design in *The Community Food Forest Handbook* provide a framework for the physicality of a food forest project and a roadmap for getting from idea to reality. And equally important is the storytelling that Catherine Bukowski and John Munsell engage in, helping readers understand how to nurture the human community that grows up along with a food forest and how to lead that group of people to create meaningful change. The ultimate power of projects like this is helping us remember our humanity through our shared daily need for food and our place in the interconnected web of people around us. But we can't get there without a solid plan. *The Community Food Forest Handbook* does much of the heavy lifting so new food forest projects have a greater chance of long-term success.

We have civilization because of agriculture. Food is, and always has been, the number one concern of humanity. Food production solutions that worked in the past won't work in the future for many reasons: climate change, overconsumption of natural resources, global conflict interrupting the supply chain, and more that we can't even predict. It is our generation's turn to figure out a future path that is reasonable and just for all. Food forests are presenting themselves as one of those incredibly smart solutions. And *The Community Food Forest Handbook* gives us a playbook to create this change together and provide a proud and enduring example for those who will come after us.

—LaManda Joy, author of *Start a Community Food Garden: The Essential Handbook*; founder of Peterson Garden Project

Introduction

The weather report called for a clear and sunny day in the low 80s. Yet it feels more like a sweltering 95°F on the sidewalk where you are standing, watching heat radiate up from the concrete. The busy road in front of you is full of cars honking and blaring loud music. There is no breeze and very little shade. You are parched and wish there was a corner store where you could buy something cold to drink. You would love to sit down, but there are no benches. Instead, you start walking. You come to a basketball court that has bleachers, but the sun-soaked metal is hot and uninviting. You keep moving, passing vacant buildings and unkempt lots—one is overflowing with abandoned cars.

On the other side of the lot, you see rich greenery. Curious, you head over to investigate. Between a vacant building and the auto salvage

FIGURE 0.1. From the sidewalk, this scene looks like an overgrown vacant lot. Look closer, though, and fruit trees, berry bushes, and flowering herbs emerge from the vegetation.

INTRODUCTION

FIGURE 0.2. This sign on the lot is overgrown with grapevines, but you can still make out that it says this space is a food forest.

yard, you find a space bursting with life. It looks like a small park, but instead of a manicured lawn, there are thick layers of vegetation and a mulched pathway heading into a jungle of trees, shrubs, flowers, and herbs. You can see cucumbers hanging off a vine and what looks like asparagus sprouting from the ground. Various plants with beautiful yellow, purple, and white flowers stand out against the green backdrop. You notice that the path splits, and one fork leads to a sign by a picnic table. Curious, you enter and sit down on the picnic bench.

Plants have grown over the sign, but some of it is still visible. You read that this place is a community food forest. You are not sure what that means, but you are glad you found it. The sign says to harvest what you want and leave some for others. As you look around and take in the scene, there is a dazzling array of textures and forms in the verdant landscape. Wandering around, you find a bench under the shade of a tree, which is bursting with ripe peaches. Picking one seems like trespassing or taking something that is not yours, but they look delicious and there are plenty. You glance around, wondering, is this a test? But the sign says harvesting the fruit is welcome, so why not?

Biting into the juicy sweetness of the fruit is a heavenly escape. You begin to relax as you become aware of chirping birds and notice

INTRODUCTION

FIGURE 0.3. Ripe Asian pears, sunflowers, and other plants provide a visual screen that partially hides the view of the adjacent auto salvage yard, demonstrating how food forest vegetation can be used to improve aesthetics.

butterflies and bees busy at work. Now that you are tuned in, these sounds of nature overtake the urban hum. This place is pulsing with life. Each flying insect has its own route, schedule, and destination. Certain leaves and plants start to stand out amid the greenery. Some plants look familiar, or at least their fruit does. A cherry tomato vine freely offering its bounty sprawls across what looks like a compost pile. Creeping thyme forms a blanket along the path. The smell of mint tickles the nose as you brush by it.

You spot a patch of sunflowers, rows of corn, and a pear tree cleverly situated to screen the view of the neighboring car lot. Underneath the tree is a mixture of herbs and knee-high plants. Nearby, ripe raspberries delicately hang from a branch reaching into a sunlit patch. You pick a pear to eat and continue meandering along the path, watching insects of many kinds pollinating flowers. You notice an unfamiliar plant that has fuzzy, soft leaves and purple flowers. Delightful smells swirl about as you step on some of the herbs underfoot.

The sensory experience full of vibrant colors, tastes, smells, and textures has replenished you. You feel relaxed, yet energized and intrigued. The sign described this place as a community food forest, but what exactly is that? It resembles a young forest, but it is unlike any community garden or park you have seen before. Why would someone create something like this and leave it open to the public? Who looks after this place? How can you learn more about it? Are there other places like this? Is there one in or near your community, and if so, how can you get involved? With these questions buzzing in your brain, you walk back out into the city street.

OBSERVING AND REFLECTING

The story above is based on our visit to the Hazelwood Food Forest in Pittsburgh, Pennsylvania. The food forest offered a much-needed reprieve in a hot, busy city. In terms of food, the yield of the five-year-old project was impressive. There were fruits and nuts—apples, peaches, pears, plums, raspberries, and hazelnuts. There were vegetables —cucumbers, asparagus, cherry tomatoes, corn, and abundant greens. Plenty of herbs filled spaces everywhere—mint, lovage, borage, thyme, and oregano. Intermixed were flowers—rose of Sharon, yarrow, asters, milkweed, and more. There were even fruiting shiitake mushroom logs in the shadiest part of the lot. Overall, the site resembled a young forest, yet upon closer inspection, it appeared to be planted with a purpose.

The Hazelwood project is a great example of what a community food forest can provide—perennial plants, green space, habitat, food, recreation, and solace all in one place. But the experience of wandering around the Hazelwood Food Forest raised questions about what it takes to create and manage such a place. What was the community backstory? How did a vacant lot on the outskirts of Pittsburgh transform into an edible refuge?

Since that visit, we have learned that community food forests are typically designed with public input, maintained by community volunteers, and available to the public for harvesting and recreational use. We were also curious about the local interests, topics, and issues at play before, during, and after planting. What we found is that there is much more to establishing a food forest than simply planting some trees, shrubs, and perennials and hoping they grow. Every community food forest site is unique. Each has a compelling story. Each requires visionaries, activists, dreamers, pragmatists, volunteers, and civic leaders. These are the people who are changing their communities and learning important lessons along the way.

WHAT IS A COMMUNITY FOOD FOREST?

The definition of a community food forest differs depending on whether the focus is on the concept or the physical space. Conceptually, to better understand the meaning embedded in the term *community food forest*, consider each word: *forest*, *food*, and *community*. *Forests* are places, but forests become social spaces through physical, cultural, environmental, and emotional connections. One such connection is *food*, which can bring people together, but others such as dialogue, wildlife, and recreation are also important. When people connect in forested spaces, they shape their social sense of place, values, and identity and create *community*. Thus, concepts of environment and society are critical pieces of a community food forest, because both influence actions like planning, planting, and maintenance that transform ideas into real projects.

An on-the-ground, physical food forest mimics the spatial and functional patterns of a naturally occurring forest ecosystem. Most people think of a forest as a large area, but from a biological standpoint a forest does not have to meet a minimum height or size requirement.

INTRODUCTION

Rather, the term *food forest* signifies a highly integrated community of plants that has various vertical and horizontal plant and root layers that provide edible products.

Modeling the structure of a young forest ecosystem in public places is ideal for two reasons. First, planting multiple perennial species in close proximity can sustainably yield food in confined

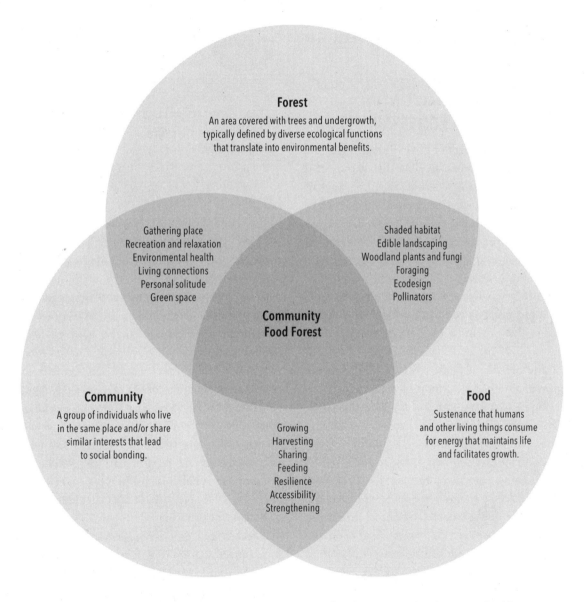

FIGURE 0.4. This Venn diagram depicts intersections between the three elements of a community food forest. Overlapping areas include the multiple functions community food forests can provide because of the relationships between the elements.

areas, which can help meet the growing global demand for nutritious production. Second, as the vegetation matures, it can create a forested parklike setting, providing a pleasing natural space where community members can gather. Even if food production in a community food forest decreases, the forest still provides valuable green space in built environments.

WHY COMMUNITY FOOD FORESTS?

More than 80 percent of the US population now resides in urban areas. This number is projected to rise in the next few decades. Finding ways to maximize use of existing open space is imperative, and increasing access to food through sustainable management of edible landscaping is one important approach among many that are underway. Community food forests have great potential in this regard because they serve multiple functions on one piece of land. This aspect makes them desirable alternatives to traditional manicured park landscapes.

The growth of food forests is similar to the growth of other local food movements such as farmers markets and community-supported agriculture (CSA) programs that serve urban areas. All are experiencing an upward trend as communities explore ways to bring food production closer to, or back into, urban areas, increase access to fresh produce, and promote knowledge about where food comes from. For example, registered CSAs were a rarity in the 1980s, but now nearly 13,000 farms market their produce through thousands of CSA programs across the United States. Farmers markets have experienced similar growth, jumping in number by over 400 percent since the mid-1990s. The relatively recent rise of community food forests

The Big Picture of Urban Food Forestry

Urban food forestry is the intentional use of perennial food-producing plants to improve the sustainability and resilience of urban communities. A wide range of activities and projects fit this definition, including community food forests, urban orchards, foraging in public urban forests, and community-wide gleaning programs. Even maintaining a small food forest on a private residential lot in a city is considered urban food forestry.

suggests they may experience something akin to this growth in the future.

One of the most powerful outcomes of a community food forest is education. A community food forest can be used to help people learn about ecology, sustainable food production, food literacy, and engagement in community initiatives, as well as introduce them to cultural practices such as foraging, gleaning, and growing perennials. It also can be a venue for social learning where people meet and take care of each other, build foundations for a shared economy, create abundance in their community, and find actionable ways to take part in forming a resilient society. (See color plate 1.)

Food literacy is a key lesson of community food forests. Community food forests introduce people to a wide variety of food-producing plants and improve their ability to identify and harvest them. Over time, this can change perceptions of acceptable and desirable food sources in a community. In community food forests, people

INTRODUCTION

FIGURE 0.5. Bees and butterflies abound on the herbs, fruits, and flowers in a community food forest, reminding people that the foraging opportunities are created for all life forms, which supports biodiversity in urban areas.

often try fruits and nuts they were completely unaware of, such as pawpaws, elderberry, aronia, and hazelnuts. Though this new awareness may seem small or insignificant, it is evidence of change. As community food forests mature and evolve, the larger and lasting goals espoused by many initiatives are likely to take root. (The appendix on page 231 lists mission statements and goals from community food forests.)

Community food forests address environmental justice when a participatory planning process is used that engages all members in site design and social activities, which define how public land should be used to serve their community. After food forests are established, they provide a natural setting for personal and community enjoyment in ways defined by those who use it. When designed with public spaces in mind, they contribute to neighborhood aesthetics by providing green space and shaded areas in which to relax as well as to engage in recreational activities such as gardening or walking. Additionally, they provide environmental services and benefits such as storm water management, biodiverse habitat (especially for birds, pollinators, and other small fauna), and healthy soil.

Community food forests contribute to civic and environmental resiliency when seed banks are planned into their purpose and trees are selected for preserving genetic material for vegetative propagation. Sites can be libraries of regionally appropriate, climate-acclimated seed and propagation material for families, homesteads, and local farmers. A community food forest at any scale can enhance ecosystem health, but it is important to consider carefully which environmental objectives are realistic to pursue given the size of a project.

SOCIAL DESIGN FOR COMMUNITY SUPPORT

Food forest design is well understood from a biological standpoint. Many websites, books,

INTRODUCTION

and other publications describe how to assess a site, design for uses of space, and group species together by function and purpose. However, their focus is typically on private property, where the forest caretaker is a single person or a family. Dynamics change at the community scale. Food forests by themselves can be complex, but they become even more so when they are community supported. The term *community food forest* is useful because it captures both the physical and social components of these projects. Community food forests serve local people while also propagating food-producing trees and shrubs, edible fungi, and medicinal plants, as well as providing environmental services and aesthetic experiences.

These projects are also a pivot point where environmental aims and the ethical and cultural foundations of human community can be combined and expressed through the design of a perennial plant community. Community food forest design must consider how plant species change over time due to natural succession. Some plants will find their niche and others will fail. Having more than one plant fulfill multiple functions in the overall design builds resiliency into the system. Similarly, there will be a natural turnover of people involved in a community food forest. The number and names of the members, organizations, and agencies may change, but transparency and documentation throughout the process is crucial for the project to survive over time. By keeping communication open and active, participants can anticipate and guide such changes. This will allow the food forest to grow positively and meet evolving conditions, needs, and values.

The manner in which a food forest is integrated into the fabric of a community is at least as important as the trees that are planted and the food that is produced. How do people come together to decide on a cohesive design? How do they care for a site in the crucial beginning years when plants are establishing? What should they think about and initially do to ensure the longevity and success of a food forest that is open to the public? A community food forest is a new model for communally growing food in the United States, and the answers to such questions are not well understood. Insight and needs vary from community to community, and we turn to key examples from our research work and our own experience running a community food forest to identify common threads.

Encouraging local and diverse involvement early on is key to the success of a community food forest. Staying flexible and working together

What Makes a Forest a Forest?

Psychological and emotional attachments to words are created through experience. The term *forest* can signify a wooded expanse managed for goods and services, as well as an area valued for its intrinsic characteristics. Perspectives vary from person to person. Some would describe a forest as a natural, lush area composed mostly of trees and shrubs. Others might think of a forest as an area primarily valued for cyclical timber production. For many, a forest symbolizes a place to explore, recreate, or simply escape. It suggests a level of wildness and messy abundance where one is surrounded by nature. For others, a forest embodies spiritual experiences and provides tangible and intangible healing.

INTRODUCTION

FIGURE 0.6. This view from June 2015 shows the gravel parking lot and plant nursery that replaced the Hazelwood Food Forest in Pittsburgh. Some remnant plants were left in place alongside the fence of the new nursery.

is central to planning and maintenance of the site, as well as to the people and relationships involved. Community food forests that embrace a grassroots approach are typically the most successful, because participants gain a sense of pride and ownership in the project. Stakeholders come together and create a shared vision, acquire land, and navigate local policy. They build sweat equity. A milestone is met when they break ground together. Another is accomplished when food is produced and community is celebrated.

Hazelwood was one of the first community food forests we visited while researching this book. One of our goals was to learn about the important community issues and values motivating the development of these projects. We also studied what role these play in shaping project planning and maintenance. Information was collected at thirty community food forests across the country. For the purposes of this book, we focused on forests open to the public and therefore did not include those located on college or university campuses. Although they are equally important, they also have very different social dynamics, with a continuous flow of new volunteers who regularly come and go. Students may even be required through coursework to participate. Community orchards consisting only of fruit trees were also omitted, because we wanted to learn from projects that intentionally mimic a forest ecosystem by planting three or more layers of vegetation. What we found is that successful projects involve complex community collaboration. They combine local expertise, insights, and interests to examine civic issues and define project goals and practices.

Our findings show that community food forests serve their immediate communities and larger town or city population in many ways, but long-term success is challenging. For instance, after visiting the Hazelwood Food Forest and meeting with site leaders, we learned that the property lease had been lost. The nonprofit community development corporation that managed the site lease did not renew the contract in favor of a local start-up business that had support from the Heinz Endowments, a major philanthropic foundation. As part of the revitalization efforts that ensued, the Hazelwood forest was replaced with a gravel parking lot and, ironically, a small plant nursery business.

WHY THIS BOOK?

Planning and adaptation are particularly important for community food forests. Exciting progress during the early stages of a project can end up in disappointment unless a long-term management plan is mapped out at the onset. Some of the most vital skills are organizing people and sustaining relationships. For example, project leaders must thoroughly reflect on community motivations, identify realistic goals, and schedule checkpoints to assess progress. However, the ways in which to best organize and lead groups of people involved with food forests has received relatively little attention. This is why this book is needed. One of its most significant contributions to the growing literature on this topic is our focus on social systems that support community food forests and other similar initiatives.

Individuals, volunteer organizations, school clubs, and the general public have helped establish community food forests across the United States. Most observable change occurs in the first year, when a site transforms from empty space into a newly planted sanctuary. When people see immediate results, the visual reward of their labor is usually satisfying enough to keep them coming back. Many volunteers also sign up to maintain a site, but a forest ecosystem matures slowly and visual change will gradually plateau. The same may happen to volunteer and even project leader interest. What happens while waiting for perennial plants to mature and produce a shareable harvest? Natural succession will make space for some trees and plants to survive, while others will die off. This can be discouraging or beget a sense of failure that can affect participation.

A community-centered approach focuses on people as one of the most important components of the food forest and is necessary for sustaining interest and participation. But how does one go about doing that? Communities are complex, and each have their own intricacies and nuances. This book provides a view into what a range of leaders and teams have faced and accomplished. We found many inspirational projects, but also uncovered some of the issues and problems that articles in the popular press do not share when covering the rise of community food forests.

The loss of the Hazelwood Food Forest after only five years is a reminder that projects can end for many reasons. Pressures on available land and community dynamics can change drastically over the lifespan of a community food forest. Being adaptive and creating plans to achieve stability well in advance of a potential threat can help reduce losses and protect property use. For example, the Mesa Harmony

INTRODUCTION

Garden, a community food forest located on church property in Santa Barbara, California, learned of investors who were interested in buying the property, which could have spelled the end of the food forest. Project leaders quickly called a stakeholder meeting to define their story, create strategies for rallying support, and discuss how potential outcomes would be handled. Such situations and responses are all part of the management needs of a community food forest, and we highlight useful real-world examples throughout this book to demonstrate how project leaders and teams have rallied and sustained community backing, addressed and overcome adversity, and cemented their food forest as a staple of their community life.

Building support for the idea of a community food forest and getting all the right players on board to acquire land and move into the establishment phase can be two of the most exciting but difficult steps. Gathering and keeping community support throughout a project is crucial to overcoming potential challenges to long-term success, such as the loss of a property lease. When community members work together and support each other through difficult moments along the journey from initial idea to groundbreaking and early establishment, strong bonds can form that have immeasurable value.

Other than the Dr. George Washington Carver Edible Park in Asheville, North Carolina, established in the late 1990s, the community food forest projects we visited were less than ten years old. Because of this, we focused much of this book on the formative aspects of place building. The early tribulations and gains of the relatively recently established sites we visited provide signposts that others can follow. Their lessons can inform and inspire us and guide projects that may be at an impasse or are just taking their first steps. Each project we studied was born out of unique circumstances, ebbed and flowed as it progressed, and accumulated a store of lessons learned and examples of creative adaptation. Our goal for this book, and the research that supports it, is to capture and share these undercurrents by telling the stories of community food forests across the country. We stitch together common themes and concepts related to history, community development, and change, along with project planning, leadership, and reflection.

You are likely interested in this book because you want to learn more about community food forests and their collaborative planning and place making: How do projects create public space that increases a community's well-being, health, sustainability, and social cohesion? When those who participate in developing a community food forest realize the impact of the project on their community, it creates pride and confidence. One of the most important outcomes of a successful project is the growth of skilled denizens who have learned what is possible when people collaborate in shaping a shared vision for their community. Of course, no one can guarantee success, but by learning from those who have gone before, we will all be better prepared to lead successful and sustainable food forest projects. The stories we tell in this book illustrate a list of community investments and management strategies that can be helpful to those who are developing and nurturing a community food forest. We want to help you prepare for and manage the social factors that can affect planning, installation, and maintenance at the community scale. As such, this book will not walk you through how to draft a planting plan for your project, nor will it lay out a universal recipe for navigating community dynamics. What it does do is summarize and organize insights and lessons learned from projects across the country and our own experience managing a community food forest.

We present concepts and frameworks that will help you make sense of the complexity involved in planning and designing a community food forest. Examples throughout the chapters help ground these ideas in real-life experiences. Longer stories of selected sites offer a rich set of community-building and management insights and practices that will help you identify the tools and strategies you need to devise a successful plan for a project in your own community. We invite you to use these stories, concepts, and frameworks as a guide for reflection and action as you design projects that nourish physical and spiritual community in the places we live.

HOW TO USE THIS BOOK

This book is divided into four sections that build off each other. Part 1 provides the foundation intended to enrich your understanding of the rest of the book, but it is possible to read other parts individually if you are directly interested in planning, designing, or building community. Four of the chapters (7, 10, 13, and 15) present in-depth case studies of community food forests selected from different regions of the country. Sharing their stories provides useful examples of how each emerged as an idea, materialized through a process, and continue to grow or exist with community support. You can read these chapters individually, but reading them in context will help anchor the concepts presented in neighboring chapters. By the time you have read through to the end of the book, we hope you will have gained the knowledge to recognize relevant patterns among the in-depth case studies and other brief food forest descriptions, as well as pinpoint unique characteristics that are worth replicating.

Part 1 covers the basics of community food forests and some of their general history. Chapter 1 provides an overview of their current state and many of the cultural and societal issues motivating their development. As we have mentioned previously in this introduction, community food forests are complex. Substantial thought is needed in their planning and management. To address this complexity, systems thinking is introduced in chapter 2 as a way to understand the many moving parts. Chapter 3 explains a framework based on seven forms of capital as exchangeable resources in the system of a community food forest. The capitals we outline represent community investments to think about when designing and managing a community food forest. This framework structures projects as integrated ecological and social systems that draw upon, and contribute to, a spectrum of community resources.

Part 2 is dedicated to various considerations that positively or negatively impact planning. First, in chapter 4, we discuss basic phases of project management that will help organize planning to take a community food forest from an idea to a planted site and community project. It also addresses considerations for adapting plans along the way based on progress toward goals and objectives. Chapter 5 delves into deeper aspects of planning for and creating change and ways to think about a project as a source of multiple community benefits and resources. An example from San Diego, California, illustrates how planning for specific change led to a community food forest that serves as an educational and propagation hub for food production, which, in turn, empowers neighborhood participants to create their own local food networks.

INTRODUCTION

Chapter 6 reaches back in time to expand on the historical roots that provide foundations for thinking more deeply about the contemporary development of food forests. By understanding the history of a place, one can also strategically design projects to work with ingrained community perspectives or use the project to create change. An example from Boston, Massachusetts, is included in the chapter to demonstrate how history can be integrated into the story of a site. We provide another example in chapter 7—the Dr. George Washington Carver Edible Park in Asheville. This community food forest is one of the oldest ones we know of on public property in the United States. We were able to acquire archived municipal documents and interviews with the original designers, allowing us to provide the first publicly available comprehensive history of the site on record.

Part 3 addresses concepts related to designing a site at the community scale. As part of this, we discuss the agroecological aspects that influence site design as well as how to think about projects as integrated systems that have social and ecological impacts through food production. A story from Iowa City, Iowa, illustrates how those concepts were front and center in designing a food forest and acquiring project funds. The language and ideas presented throughout chapter 8 can be seen as groundwork for understanding how to package and communicate a community food forest idea when partnering with professional allies. Chapter 9 has two important messages. The first part of the chapter stresses the importance of thinking about community food forests as public space and designing them to attract community interaction. The second part introduces groups of professionals who can be allies when designing a food forest and managing urban environments safely and efficiently.

Chapter 10 describes a site in Basalt, Colorado, where the intimacy of a small-town setting and tight collaboration among local agencies, public services, and community members led to quick transformation of a park into a food forest and seed library.

Part 4 focuses on the concept of community, which is such an important part of these projects. In chapter 11, we dissect what community means in different contexts to help define the audience a public food forest project can attract and serve. In chapter 12, we revisit the idea of systems thinking introduced in Part 1 and relate it to the intentional creation of social systems that support a community food forest project. Chapter 13 is the story of the Beacon Food Forest in Seattle, Washington, one of the most publicized community food forests in the country, which sparked numerous conversations on the limits and potentials of projects open to the public. The site is scheduled to eventually cover seven acres, demonstrating that perseverance can have big payoffs and that a larger scale is possible when a project has strong community support.

Chapter 14 discusses concepts of leadership within the context of community food forests. Collaboration and governance structures are discussed in relation to what we have seen work at a variety of sites. The final chapter of Part 4 tells the story of a site in Bloomington, Indiana, as a way of demonstrating how a project can develop over time and give birth to a community-based organization of volunteers who learn how to work together.

In the book's conclusion, we present all the major lessons learned from the case studies. The growing number of projects represent a wealth of valuable insights. However, cross-project communication and networking between

INTRODUCTION

community food forests in the United States are lacking. The dearth of inter-site communication inhibits the emergence of a shared vision of change and makes the growth of community food forests difficult to classify as a social movement. Sites across the country are gaining a great deal of attention. Taken together, they loosely exhibit coordinated social activism, but the lack of networking or formal connection is a missed opportunity for sharing insights, failures, and successful strategies that could accelerate how new community food forests are designed and improve how existing ones are managed. We touch upon potential solutions to those current barriers, so community food forests can serve a more concrete role in urban planning and local food systems. In addition, we highlight current research contributing to the development of urban food forestry, which encompasses community food forests.

Overall, our aim is to share case studies and concepts that will help you establish and sustain meaningful community food forests. We have thoughtfully explored the challenges of leading a complex project that is open to public participation and distilled significant findings into these pages. We supply stories dispersed throughout the book as examples and insights of what we saw at successful sites or ones that faced challenges. However, we also stress that this handbook is a starting point on your journey toward creating a long-lasting community food forest. The more we learned about the intricacies involved with planning and designing food forests for communities, the more we realized we would not be able to cover it all in one book. We encourage you to visit the website Catherine has developed, www.communityfoodforests.com, for listings of other important books, articles, and media pieces, along with links to other websites and digital assets.

PART 1

Understanding Community Food Forests

CHAPTER 1

Community Food Forests on the Rise

Community food forests are capturing the imagination of people in neighborhoods, towns, and cities across the United States. Their popularity reflects a value shift in urban cultural pockets. The message is a desire for public space, where possible, to be ecologically designed with perennial and annual plants that produce food and herbal medicine, enhance nutrition, promote food literacy, and provide a useful and safe place to gather, recreate, and work together. This is all while engaging people in active participation to create the places they want to live in and to voice their opinion through action. By developing these spaces, people are stating that ecologically healthy green spaces and sustainable local food production are valued, especially in the face of urban population growth. Communities will innovate, using all the resources they can harness, to increase the presence and quality of such resources in urban landscapes.

Community food forests also serve a deeper purpose by helping community members form bonds through collective labor and learning. Participants often discover shared interests such as local and foraged food, social justice, environmental stewardship, resiliency, and self-sufficiency. Uniting around common causes, people invest in and build diverse assets in their community and this personal development and civic collaboration benefits society. Questions emerge on why we feel disconnected from land and how to develop the culture of sharing abundance, human skills, and knowledge needed for survival in the modern world.

Many communities today embrace the belief that local food should be readily available, and that much of it could come from within city or town limits using ecologically sustainable design and safe urban production methods. The reinvigoration of this form of community spirit has helped focus a new urban agriculture agenda. Community food forests are strongly linked to local food, food justice, and civic agriculture movements. Participation in a community food forest project can lead to critical reflection on our current agricultural system and urban landscapes. Typically it motivates people to work on influencing political action and policies.

Community food forests raise important questions about access to fruit trees and other edible perennials in public places. They introduce people to foraging for "wild plants"—edible and herbal species—in public parks, forests, and rights-of-way or to gleaning unharvested produce to supplement community supply.[1] These issues are increasingly observable in the public agenda in terms of sustainability, food security, environmental justice, and urban green infrastructure.

Community food forests can be found in a variety of places. Churches, universities, and intentional communities have planted food forests on their campuses. They are increasingly found on public property managed by public works agencies or parks and recreation departments. Regardless of where they are located, these projects are open to the public. Volunteers and civic organizations are often involved in their development and oversight. Enthusiastic faculty and students tend community food forests on university property. Dedicated groups of congregational volunteers encourage and guide member participation in projects coordinated by churches. On public grounds, the collaboration and communication between agency employees, project leaders, and volunteers is essential for effective management and community support.

Community food forests are part of a cultural transition and represent local efforts to build abundance and share opportunity. Even more important, they can contribute to meaningful personal, civic, and ecological stewardship that often is lacking in our lightning-fast, digitally driven, consumerist lifestyles. Community food forests offer a way of experiencing the direction in which this shift is taking us. It is this possibility of deeper meaning in our lives that makes community food forests such a compelling and inspiring movement.

SOCIETAL SHIFTS

To our knowledge the first contemporary community food forest launched in 1997 in Asheville, North Carolina. In chapter 4 we share the story of how the vision of two young college graduates interested in permaculture and sustainable cities transformed an unused parks and recreation property into a contemporary edible public park. The New Urbanism movement of the early 1990s, which promoted walkable cities with environmentally friendly habitats, influenced these young people to find the resources needed to make their dream a reality. The edible park has undergone many changes since its installation in 1998, but twenty years on, it still produces free fruit, nuts, and herbs while serving as a place for learning and sharing.

About a decade later a couple of community food forests arose in progressive cities, but it was not until around 2012 that the extensive media coverage of the Beacon Hill Food Forest in Seattle catapulted community food forests into the public eye, and the rate of new projects began to accelerate. As of 2018, we are aware of more than seventy community food forests in public spaces in the United States in communities of all sizes, spanning from the Pacific Northwest to the Deep South. Figure 1.1 on page 19 shows what we know about community food forest locations in the United States at the time of publication. You can see they have sprung up across the country, demonstrating that food forests are possible and relevant in any community.

The rise of community food forests correlates with larger changes in social consciousness and dialogue about local food production and access. Green space that provides healthy and fresh food to people and important environmental services to communities are some of the most compelling benefits of food forests. Planning for the Beacon Hill project (see chapter 13 for the story of this food forest) began in 2009, the same year that First Lady Michelle Obama initiated a national conversation about food production. At

FIGURE 1.1. There are more than seventy community food forest projects in communities throughout the United States and several in Canada, too (not pictured here). More are announced each year, as people learn about this concept or are inspired by reading about food forests in other places. We want to create a comprehensive list of community food forest sites. If you know of a community food forest not shown here, visit www.communityfoodforests.com to enter the information.

that time, dialogue about the need for new food systems and community building was increasing. Reclaiming sovereignty over food production using environmentally dynamic systems to do so has been central to many contemporary community food forest projects.

Food is embedded in culture, politics, health, and community development. Growing food is empowering, and its fundamental place in humanity transcends ethnicity, nationality, race, religion, sex, socioeconomic status, and political affiliation. Many people have lost connection to traditional knowledge and customs, but in recent years, wild foraging and historical food preparation have experienced a revival. Communities are re-empowering their control over food, and formerly detached consumers are becoming informed local producers and advocates. Community food forests play a role in this awakening by enabling exploration of alternative food sources.

COMMUNITY FOOD FOREST ROOTS

Creating public spaces that yield perennial produce for communities is nothing new. Many neighborhoods and organizations have intentionally planted community orchards or edible landscaping for public use. Village Homes in Davis, California, developed throughout the 1970s, is one example where the designer used concepts of community sustainability and permaculture to incorporate edible landscaping into a seventy-acre suburban neighborhood. Perennial edible vegetation is still found along walkways between homes and in public spaces, including orchards, an almond grove, vineyards, and community gardens. When strolling along the paths of Village Homes, it is easy to pick grapes, figs, plums, pomegranates, mulberries, and grapefruit in the span of a thirty-minute walk.

In the 1990s, the Serenbe Community in Atlanta, much like Village Homes, incorporated edible perennial landscaping along with naturally conserved open land, forests, and zones for agricultural production. A more recent example of this type of intentional design is the twenty-seven-acre Community First! Village in Austin, Texas. The development provides affordable housing and job support to the homeless and is built around Genesis Gardens,

FIGURE 1.2. This park in the Los Angeles area features an art installation of fruit trees, a creative variation on the concept of a community food forest. This fenced area represents the "eye" or central part of the installation; other fruit trees are dispersed throughout the park and surrounding neighborhood.

an organic farm with 200 fruit and nut trees maintained by residents and volunteers.

These intentional communities and many like them share characteristics with community food forests. A significant difference is that production areas were created for use by members of the intentional community (in some of these communities, though, people often freely wander in to admire the landscape and harvest products). Most community food forests, on the other hand, are intentionally open to public harvest and rely on volunteer maintenance. Many are used as an educational tool for promoting what is horticulturally possible, because most are not at a scale to provide sufficient harvest to feed an entire community. Often, one goal is motivating volunteers and others involved to transfer the knowledge and skills gained on site to their private residence and other neighborhoods.

Organizations such as Fallen Fruit (Los Angeles), Portland Fruit Tree Project (Oregon), the Philadelphia Orchard Project, and the Boston Tree Party are leaders in urban orchard establishment and promote gleaning from existing trees in their municipalities. These organizations also support participatory public art projects of "civic fruit." The work of the organization Fallen Fruit, in particular, provides an excellent example of how policy planning and creativity can benefit a community food forest. The city of Los Angeles, where Fallen Fruit works, could not support fruit tree planting in public space primarily because of maintenance and liability issues. Fallen Fruit proposed that their community food forest project use fruit trees as an "art medium." In this way, they were able to create a community food forest in a public park with funding from the City Arts Commission.

A NEW MODEL

Along with community food forests, urban orchards and civic fruit installations are part of a larger pulse to increase perennial food production in community space. They represent a renewed public interest in resilient agriculture and sustainable production that emerged in the second half of the 1970s. Environmental policies and liberal ideas in the 1960s shaped the minds of many Baby Boomers, some of whom have become today's community food forest leaders. Their perspective and leadership toward these ideals has profoundly influenced many Gen Xers, as well as their Millennial children.

Paralleling the social consciousness of earlier generations were programs and publications that set the stage for current awareness and interest in local food production. In 1976, the USDA helped coordinate the Urban Gardening Program by establishing offices in cities to promote community gardening. Three years later, the American Community Gardening Association was formed. Books about sustainable agriculture exploded onto the scene, most written by men who grew up during rapid suburban expansion in the mid-twentieth century. They struggled to determine what came next in light of their experience and the stories told by their parents and grandparents about the transition from a mostly agrarian society to an urban and mechanized nation.

At the same time, polyculture food production systems were emerging, including agroecology, agroforestry, and permaculture. Agroecology is a holistic land management framework that applies ecological principles when producing food and aims to improve food literacy and social justice. Agroforestry is a set of agroecological practices that intentionally add trees to agricultural land or cultivate woodland crops such as ginseng or mushrooms in woodlands. Permaculture, which is another agroecological system, also emerged during this time through the leadership of Bill Mollison and David Holmgren. Permaculture is a worldview and approach to land use planning and management that considers nature to be the teacher and humans the students. Agroforestry was promoted through research and extension services, first in developing countries and later in industrialized ones. Permaculture, on the other hand, entered the informal public domain and won over dedicated followers with its commitment to holistic and spiritual land use.

In the 1980s, agroecology was formally recognized, and the inaugural International Permaculture Conference and North American Agroforestry Conference both occurred. The Slow Food movement began in reaction to fast food chain outlets appearing in Italy, among other reasons, and now has a presence in more than 132 countries.[2] Food forests align well with Slow Food's philosophy of ethical consumption and focus on decelerating our lives, being more intentional, and enjoying food in all phases from growing to harvest, cooking, and eating. Slow Food adherents believe in achieving desirable outcomes over time, through continuous adaptation and experimentation, and the movement's principles fit well with the production cycles of woody edible perennials that can take time to bear fruit.

LOCAL FOOD IN THE NEW MILLENNIUM

As the 1980s carried over into the 1990s, the United States Congress passed the Urban and Community Forestry Assistance Act to support urban forest management for environmental and social goals. The USDA National Agroforestry Center, the Association for Temperate Agroforestry, the Center for Agroforestry at the University of Missouri, and the Alliance for Community Trees were founded during this time. These institutions focused attention on the importance of trees serving multiple functions in production systems. Also during this period, Robert Hart wrote the influential book *Forest Gardening*, which describes his experience growing edible perennial plants in his backyard in England. Hart's property and writings inspired many others to establish food forests.

Smart Growth initiatives for planning sustainable cities and healing fragmented communities became increasingly popular. The field of sustainability, especially in terms of international development, also gained prevalence. The first agroecology textbook was published in 1997, and the local food system movement picked up momentum in the new millennium.[3] The Community Agroecology Network was formed and the first World Congress of Agroforestry met in Orlando, Florida. Permaculture educators Rob Hopkins and Naresh Giangrande started the Transition Town movement in Totnes, England in 2006. First, they asked their students to create an energy plan for a community in the face of peak oil. Then, Hopkins and Giangrande refined and replicated the results to integrate permaculture practices into community design and management.

Modern sustainability movements such as Transition have many goals. These include lower energy consumption, improved community resilience, environmental rehabilitation, and designs of new ways of living that are nourishing, life-affirming, and socially meaningful. The food forest concept fits well with these objectives because food forests contribute to community resilience by promoting self-sufficiency, creating green space, and providing a local source of regionally acclimated food crops.

The 2000s saw the publication of multiple permaculture, forest gardening, and holistic farming books, including Barbara Kingsolver's highly popular *Animal, Vegetable, Miracle*, which helped galvanize the locavore movement.[4] The White House planted an organic vegetable garden to highlight the link between gardening and nutritious food. However, what set things in motion among the public consciousness was not simply a growing appreciation of local food, but the catastrophic collapse of the US economy in 2008.

THE RISE OF COMMUNITY FOOD FORESTS

During the Great Recession of 2008, the notion of raising food as a community offered a way for people to envision being both productive and cooperative in the midst of a shared struggle.[5] A broad range of people faced issues of food insecurity and lack of access to affordable, nutritious food following nationwide economic upheaval. Social movements such as the Occupy Movement organized people around their distrust of Wall Street, big business, and government.

In the midst of their fear and disappointment, people bonded during discussions about community issues. They also brainstormed local solutions and actionable steps to create the changes they thought would build a better future. In the process, they formed a type of social capital that would serve as a foundation for some of the earliest community food forest projects. Around this time, there were very few active community food forests in the United States, but planning for many more soon began.

For example, a group known as Occupy Vacant Lots arose out of the Occupy Movement in Philadelphia and made turning abandoned areas into food forests their mission.[6] The momentum has only increased following the Great Recession. At least eight community food forests were established between 2010 and 2011. From 2012 to 2014, at least twenty more began. By the time of this writing, there were some seventy, if not more, new projects across the United States.

The rapid growth in community food forests suggests that a new environmental and agricultural consciousness is taking root. A common thread running through all of them is a desire to share topics and knowledge that were once central to community life but were in large part lost during the technological and social transitions of the last century. An ecologically grounded approach to agriculture includes an understanding of how to use natural principles for holistic crop production. When combined with an intimate knowledge of local resources, conditions, and techniques, these topics are inseparable from community food forests. We will illustrate these concepts throughout the book and emphasize their role in shaping the future of local food production and community cohesion through the stories of community food forests from regions across the United States.

UNDERSTANDING POSSIBILITIES AND SHAPING THE FUTURE

Today's excitement about local food and nutrition can be a powerful motivator that spurs people to try something new like joining a community food forest project. Their positive energy is important, but it needs to be guided. On its own, it is not enough to sustain a community food forest project. Project leaders looking to help shape their community and its collective work can use systems thinking as a framework when designing and implementing a community food forest. A necessary part of this is defining the values associated with local food so as to motivate community members to take part in civic activities. In this way, an array of stakeholders, including nonprofit organizations, academic institutions, local government, and community-based organizations, can help shape their community.

Systems thinking has evolved over time into a holistic and adaptive framework. It lays out a process for pursuing a goal while keeping the relationships between people and their environment at the forefront. Systems thinking helps identify ways to effectively work together as a group of people who want to create change in a community by developing a community food forest. Systems thinkers are better able to see community connections that are not always clear. Thus, they can more effectively find ways to achieve their goals and create the change they seek.

Working effectively with the physical attributes of a site while adequately serving a community in the design and management of a community food forest is no small task. It requires trade-offs between ecological design and meaningful community space. One cannot override the other, and because of this, systems thinking is both helpful and warranted. Food forest leaders must balance plantings and design at a given site with the activities and services people want in that particular place. And they must consider this balance in terms of system equilibrium when everything is calm as well as during times of change. Today, green space is not just "a little bit of nature" in the cityscape. Rather, it encompasses multifunctional systems that yield lasting payoffs such as leisure, health, and well-being; food production; habitat conservation; watershed management; urban canopy; and social interaction. Community food forests encourage people to think in terms of abundance and inspire neighborhoods to act on their own behalf. They promote a culture of sharing, stewardship, and nature-centered health.

Our goal in this work is to share real accounts of community food forests and tell stories about why projects started, how they lined up partnerships and built support, and the ways in which planning and implementation happened. We also discuss how maintenance was handled and cover the ups and downs of these initiatives. Alongside these reports, we also present key concepts that help frame and translate these anecdotes in a way that you can use in your own community. Our hope is that through our descriptions and accompanying analysis you will grasp the reasons people work on community food forests, how systematic planning and community perspectives play a central role in project success, and that people are as important as the plants when it comes to community food forests. We also hope you will gain, like we have, a better understanding of people's experiences working on food forest projects inspired by making a difference in their community.

CHAPTER 2

Systems Thinking for Community Food Forests

Systems are all around and within us. They exist in various sizes with different degrees of complexity. They are as small as a single cell and as big as our solar system. The shape and size of some, such as an ecosystem or community, are difficult to determine. Other systems—such as a house, a school, or the human body—are more tangible and have defined limits. Given that, what is the definition of a system? A system is a group of connected and interdependent elements meaningfully organized to function in a particular way or achieve a purpose. The elements are connected through various relationships with each other and their environment. Over time, these connections produce patterns of behavior, which create the structure of the system. The three main ingredients that create the structure of a system are its *function* or *purpose*, the *elements* that make up the system, and the *connection* between those elements. We use this framework to explain how to think about designing and managing food forests at the community scale.

A system, regardless of size, exists as its own unique entity while simultaneously connected to or nested within other systems. The cells of our bodies are a great example of this concept. Each independent cell is connected to other cells, all of which are located within a human body. A garden is part of a larger ecosystem. Similarly, a group of people working on a public food forest are part of the community where they live. Identifying and studying systems is an important task for community food forest project leaders because there are numerous interrelationships. Leaders must anticipate how all aspects may be connected, how they change over time due to internal shifts, or how they are affected by changes in other systems. This chapter covers systems thinking and how it relates to the duality of ecology and society inherent in community food forests.

THE VALUE OF SYSTEMS CONCEPTS

A system organizes elements to achieve something, and community food forests are typically created to achieve multiple objectives. The purpose or function of a system is determined through its behavior. Donella Meadows tells us that "The word *function* is generally used for a nonhuman system, the word *purpose* for a human one, but the distinction is not absolute since many systems have both human and nonhuman elements."[1] This book is divided into parts that will help you think about elements in a community food forest, both tangible (e.g., vegetation, infrastructure, and people) and conceptual (e.g., concepts of community and certain forms of community capital).

It provides strategies and tips for planning how to include elements in a way that links them to community food forest goals. Throughout the book we encourage you to think about how to intentionally design connections between elements in your community or create the conditions that allow specific relationships to form.

Systems thinking underlies much of the rest of this book, and in this chapter we introduce some of its main concepts and how they relate to community food forests. We ask that as you read this book, you think about development of community food forests from a systems standpoint to help make sense of the many social and environmental elements that are interlinked and affect each other. Community food forests are deeply intertwined ecological and social systems. Individuals connect to organize a community group defined by a common cause with a sense of purpose rooted in creating a food forest. As a result, they form an ecologically designed place with social relevance. That, in turn, attracts more individuals to join the community group, which increases the capacity to expand the scale of the food forest itself.

Connections Between Elements

Interrelationships shape our world and are key parts of systems. Changes in one element inevitably affect another. Relationships between elements in a system are often harder to see than the physical parts, but they influence how complex systems behave. In a food forest, it is not simply assembling all of the right elements such as plants, soil, compost, and water that creates an ecosystem. More significant are the relationships between those elements that create behaviors such as nutrient cycling or plant regeneration.

The well-known saying "the whole is greater than the sum of its parts" refers to a systems characteristic best described as "emergence" or "synergy." In other words, behaviors that cannot be explained or attributed to any singular part emerge only when all the parts function together. Synergy arises from the relationships between the parts, rather than from the parts themselves. Therefore, simply grouping a bunch of parts together does not necessarily mean they will function as a whole. Relationships between the parts must be cultivated and maintained.

A bicycle is a good example. Lay out all the parts of a bicycle side by side and they each exhibit their own unique, independent characteristics. These individual pieces are interesting on their own, but until they are assembled, they will not function as a bicycle. When an external force (a human pedaling the bicycle) acts on the bicycle, it creates synergy and motion. The ability to transport a human from point A to point B emerges from all the parts working together.

Nested Systems

Charles and Ray Eames's 1977 short film, *Powers of Ten*, vividly demonstrates the interrelationship between systems at various scales. At one point in the film, the camera pans out from a picnic scene by factors of ten until achieving a view atop the universe. Then it drills down to the atomic scale inside one of the humans enjoying the picnic. The experience of this short piece is both overwhelming and empowering. The magnitude of relationships between systems of different scales makes us seem at once inconsequential and also significant.

Natural Systems

Food forests are designed with polycultures that integrate diverse species into an overall space. Natural ecosystems are great teachers about polycultures and how they function. Understanding them, even on a basic level, guides decisions on placing and managing different elements of a food forest in relation to each other. Intricate interdependencies need to form to create a low-maintenance system that helps desirable species thrive. Thinking in terms of systems helps one grasp and make sense of the complexity involved in designing the physical space of a food forest.

Carefully selecting species that have overlapping functions creates redundancy. This is beneficial, because even if one species fails, functional ecosystem relationships are maintained. Larger system patterns, such as nutrient cycling, adapt and persist. Likewise, including people who have overlapping functions, skills, and abilities in the support network of a community food forest builds resiliency, allowing the food forest to thrive even in the event that a person leaves the project. As people form a relationship with a community food forest, they reconnect as part of a natural ecosystem. Functions and behaviors such as environmental awareness and food literacy often emerge as a result. These outcomes benefit community well-being and environmental stewardship.

In the community food forest context, in any given place individuals exist who care about issues such as food access or environmental conservation, but they may not be connected. A flyer or local newspaper ad announces the idea of creating a community food forest and attracts those people to the same room for a meeting. The external force, which in this case is a common cause, acts as a catalyst to assemble the parts (the stakeholders) and stimulate connections to form between them while fostering synergy to create a community food forest.

Function or Purpose

The intended outcome of a system will determine how it is designed. Using the bicycle example again, there are design differences between types of bicycles (e.g., mountain versus road racing). The preference for a specific type of experience a person wants will influence which design they are attracted to. Applying this idea to public space, certain types of people will be more attracted to a community food forest rather than a traditional community garden. As people shape the project, their preferences can affect how the system functions and which community benefits emerge. If there is a preference for supporting pollinator species or hosting community potlucks, the community food forest should be designed to support those interests.

A community food forest is designed to be a healthy ecosystem and active social system, both of which are complex. When complex systems have problems, those problems also tend to be complex. Conventional problem-solving methods, which typically use linear thinking, focus on dividing a system into discrete parts. The thought is that a solution is best determined if each part can be better understood and optimized. This seems logical, but unfortunately, this type of problem solving often does not lead to expected results when it comes to community food forests because it fails to address dynamic relationships. It is important to recognize that an action today does not immediately, or even in the long term, have a linear pathway to the desired result. This does not mean, however, that the solution to a complex problem also needs to be complex. We certainly learned from many projects that small changes can have large impacts.

People sometimes characterize a system as "broken." In actuality, networks and connections are always evolving, but they may not function in a way that produces desirable results. The same is true for a community food forest. The ecosystem and social system are always working, but parts or relationships might require tweaking. In that regard, the emphasis should be on changing relationships to improve outcomes. Negative feedback can sound the alarm for change while positive feedback helps balance and amplify desirable outcomes.

Feedback Loops

Feedback loops are helpful for adjusting system performance to meet desired output or goals. The input, whether positive or negative, is used to regulate or modify system output until it functions as intended. In our research, the need for flexibility emerged as a pattern across sites. Embracing flexibility and adaptive management strategies is very helpful when handling unexpected outcomes and responding to feedback. Many community food forest leaders shared that this was one of the biggest lessons they learned.

For example, it may take multiple public planning workshops to design a comprehensive food forest complete with path layouts, plant arrangements, and more. One workshop format often used by designers and planners to develop a vision for a project and gather feedback from stakeholders is known as a charrette. During a charrette, project concepts are explained to participants and then their ideas, values, needs, and concerns are collected through open discussion. As much of the feedback as possible is integrated into the design, which is presented again to the attendees. New feedback on the advanced design is again collected and incorporated. Charrettes are a way to create a feedback loop when designing community food forests. The process is intensive, requiring patience and flexibility. Community food forest designers must plan for the ever-changing interaction between perennial plants, structural features, and people. In the end, the plan on paper rarely transfers perfectly to the physical realities of the site, but it brings to light community design preferences. Thus, it is best to think about your community food forest plan as a compass that will help point your project in the right direction.

Being prepared to adapt to change as a community food forest evolves increases resiliency and therefore improves chances for permanence. Adaptability will also help you design evaluation metrics, which should include regular surveys of volunteers and users of the food forest. Identifying metrics that provide feedback on the effectiveness of a community food forest provides direction for adaptively managing both plants and people. It is worthwhile to conduct roundtable meetings with leaders to gather feedback and discuss incorporating ideas into future steps for the project.

If feedback indicates a need for change, the question of what to change and exactly how then arises. Defining leverage points—places within a system where the most change can occur with the least amount of effort—helps define what type of adaptation is needed. When considering leverage points, it is helpful to weigh information about the different parts of a system. For example, perhaps most people in the community work long hours and have children. In such a scenario, poor attendance at meetings could be more a matter of availability than interest. Leverage points might be the timing of the meeting and whether children are allowed to attend. A productive change might be to use freely available technology to broadcast the meeting over the Internet to participants who cannot attend. Or the number of monthly meetings could be increased, with the same material covered at each meeting. Expectations of how many people will attend may also need to be adjusted. Other solutions might be to provide child care during the meeting or to change the meeting format to include dinner.

NESTED SYSTEMS

When all the parts of a food forest synergistically work together, they become a self-sustaining system that can produce food year after year. Its self-sustaining characteristic makes the food forest an attractive model because most people equate such biological autonomy with something that is very environmentally friendly, while also requiring less maintenance over time. Having a food forest nested within the community affords more opportunity for residents to become self-sustaining through access to food and the presence of a common goal on which to build relationships. Cultivating the human relationships within the system increases the chances of producing self-regulating, low-maintenance biological outcomes.

The Food Forest (Eco)System

Forests are self-regulating (eco)systems that constantly adapt in response to complex feedback loops. A forest ecosystem consists of plants occupying niche spaces in multiple layers in pursuit of nutrients, water, and sun. The natural arrangement of plants in a forest creates a systems-level strategy to prevent the catastrophic spread of plant disease, to provide seasonal nutrients for temporal flowering of plants, and to help control populations of damaging insects. It also helps accumulate organic matter and soil nutrients, via specifically enabled nitrogen-fixing plants and leaf litter decomposition.[2] The patterning of plants and fungi in a forest, and the way it changes over time, creates a redundancy of functional roles that help sustain a level of resiliency. This enables the forest to overcome challenges such as species-specific blight or temporary drought, while also providing innumerable environmental services.

In humid tropical climates, light and water are available year-round. Natural succession proceeds at a quicker pace than in temperate climates. As plants vie for light and nutrients, they adapt often to the highly dynamic setting, and forest vegetation layers establish quickly. Food forests are often modeled on a template of seven or eight vegetation layers, which originates from a tropical forest structure. We list examples of these potential food forest layers in table 2.1, along with some illustrative plants to show a range of possibilities.

UNDERSTANDING COMMUNITY FOOD FORESTS

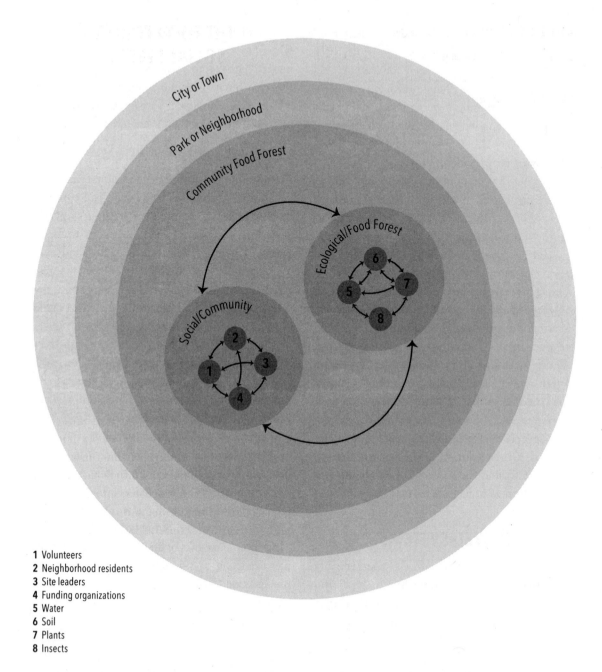

1 Volunteers
2 Neighborhood residents
3 Site leaders
4 Funding organizations
5 Water
6 Soil
7 Plants
8 Insects

FIGURE 2.1. This diagram is an example of nested systems in a community food forest. The community contains volunteers, a core group of leaders, neighborhood participants, and donors, all of which are connected to each other. They influence the community system but can also be affected by it. The food forest contains trees, herbs, water, and soil, which affect the food forest and vice versa. The community food forest is located within a park system that may have constraints on the food forest and may change how the park is used.

TABLE 2.1. Food Forest Vegetation Layers

Layer	Representative Plants
Canopy trees	Oak, chestnut, pecan, mulberry
Subcanopy trees	Apples, pears, peaches, plums, hazelnuts, pawpaw
Bushes	Raspberries, blackberries, currants, cranberries, gooseberries
Shrubs	Comfrey, rhubarb, asparagus
Herbs, flowers	Sunflowers, rosemary, basil, sage, yarrow, lovage
Ground covers	Squash, lamb's ear, strawberries
Roots	Jerusalem artichokes, onions, potatoes, wasabi
Fungi	Shiitake, maitake, lion's mane, chicken of the woods

In temperate forests, succession takes longer because of the shorter growing season. As a result, the length of time it takes for a food forest to reach a more stable condition is more variable. The types and number of species it can support depend on many factors such as climate, site conditions, and the availability, source, and form of planting stock such as native and nonnative seeds, cuttings, or transplants. It is important to keep in mind that it is not necessary to include all layers listed in table 2.1 in a community food forest. In temperate or drier conditions, there will typically be fewer.

Our experience visiting sites around the country indicate that two or three intentional layers of vegetation are most common at the beginning of a project. Often, more layers are established in successive phases. When a food forest matures, and its canopy fills in, the highly shaded understory will naturally reduce the number of species due to light limitations. This can open space for other uses. In general, decisions about density during establishment and throughout the life of a community food forest are based not only on site condition and forest mimicry, but also on the preferences of community members and project leaders as well as on local policy requirements.

THE FOOD FOREST SOCIAL SYSTEM

Relationships between individuals and/or institutions define social systems. Relationship patterns form roles, status, and characteristics of a group, all of which contribute to its structure and function. The people who design, install, and maintain a community food forest affect it just as much as, if not more than, the ecosystem dynamics. Thinking in terms of systems provides a framework for intentionally constructing the social system or community group that will support the food forest. We delve deeper into this topic in chapter 12. For now, it is important to remember that the number of stakeholders in a social system is less important than the relationships among them. The strength of collaboration through thick and thin, foreseeable and unforeseeable change, is the critical litmus test.

Just like the natural succession that occurs in plant populations as site conditions change over time, the people and organizations involved with a community food forest will similarly evolve and change. However, it is often difficult to predict how people will react to changes. Including people and organizations that perform many different, and sometimes overlapping, functions will increase the resiliency of a project when change is encountered. Recognizing that community food forests exist within this complex variety of nested human systems helps leaders and volunteers navigate the ebbs and flows of human activity.

One example that helps demonstrate interrelationships between people working on a community food forest is that of the public-private partnership, such as a community food forest located in a public park. A park is a system in and of itself, with various subsystems and elements, such as basketball courts, walking trails, and swing sets for various functions like recreation or leisure. The park is typically managed by a parks and recreation agency, which usually has an office staff with its own culture and bureaucracy. The agency is part of a larger municipal governance that is linked to the county, to the state, and so forth. It is a complex web of nested systems.

A group of private citizens and organized nongovernmental organizations who seek to develop a community food forest on public land must consider how their own culture matches that of the agency responsible for the land. Using a systems lens, here are some points and patterns to consider when balancing public and private community participants and partners:

- What are their perspectives on land use?
- What structures do they have in place for managing their programs?
- What types of projects are they involved in?
- How do they maintain properties?
- What types of concerns do they commonly have about new projects?

At almost every community food forest we visited, we found that maintenance and liability are major concerns. Paying heed to this is important when determining how best to approach a landowner or property manager to discuss a prospective food forest project. Immediately addressing such concerns shows consideration and opens the door for collaboration.

STOCKS AND FLOWS OF COMMUNITY CAPITALS

Models, diagrams, and tools exist that can help you conceptualize and clearly describe the whole system involved in your community food forest project: to identify its structure, visualize how it functions, and determine leverage points. You can harness these tools to develop a story about the type of community food forest system you want to create. For example, when working on a project that is not producing desired results, it is easy to place blame on others. Telling a systems story prompts people to identify all the characters in the story and how they contribute to the plot. Recognizing everyone's role in this way spreads the responsibility for creating change and helps avoid the tendency to place blame on others.

A good way of thinking about community food forests and change is to use the concept of stocks and flows. In her well-known primer on systems thinking, Donella Meadows explains stocks and flows using a bathtub metaphor.[3] A stock is as an asset or resource that accumulates or decreases by flowing either in or out, like water in a bathtub. Pipes are the mechanism through which inflows occur. An open or closed faucet controls the rate of inflow. Outflows occur in various ways: a drain, a crack, or even water spilling over because the tub becomes too full. In order for the water to remain at a consistent level, the system must behave in such a way that the inflow and outflow remains constant and balanced. From the standpoint of a community food forest, there are stocks that regulate forms of development relative to civic investments. These human-centered stocks are often referred to as *capitals*.

The word *capital* typically conjures up thoughts of finances or the monetary economy. So do the terms *assets* and *stocks*. But in its purest sense, capital is any mechanism, resource, or asset that improves performance—it does not have to be monetary. Capitals flow through a system,

TABLE 2.2. Community Capitals Comparison

Community Capital Framework	Permaculture Flower of Ethics and Design Principles*	Ethan Roland's 8 Forms of Capital†	Definition	Examples of Relationships to Community Food Forests
Natural	Land and Nature Stewardship	Living	Living and nonliving components of an ecosystem that together form assets that support life.	Creating a natural green space that produces food and provides key on- and off-site environmental benefits such as increased vegetation, diverse habitat, better water and air quality, and improved aesthetics.
Human	Health and Spiritual Well-Being; Education and Culture	Intellectual; Spiritual; Experiential; Social	People and their beliefs, aspirations, energy, skills, education, and ideas that together help communities function and thrive.	Engaging and organizing community members to maximize use of their collective skill sets toward useful green infrastructure and food production projects such as gardening, fencing, plot planning, and plant design as well as organizing others and creating governance structures. Learning about alternative food sources, planting methods, creating community change, and environmental stewardship.
Social	Education and Culture; Finances and Economics	Social	Strength, direction, and focus of associations between people in a community and these associations' roles in collective identity and networking.	Building networks of citizens from various backgrounds and interests who come together to bond over similarities or bridge differences and engage with each other while participating in community greening and food production activities.
Cultural	Education and Culture	Cultural	Preferences and perceptions of people about their community in terms of its norms and values.	Enhancing community concepts of food and green infrastructure by producing in underused or strategic locations to improve public access, introduce new and innovative methods, and demonstrate green infrastructure techniques.
Political	Land Tenure and Community Governance		Formal and informal organizational structures that allocate resources and drive policies and objectives.	Residents learn community governance strategies to create change in communities and formally work with, and receive support from, government agencies or institutional organizations. Organizations and community groups learn to work together toward safe, responsible, and legal establishment of local food production, green space, and legislative change.
Financial	Finances and Economics	Financial	Monetary assets and savings in a community that support households and increase goods and services.	Leveraging progress to secure support from external donors or investors, followed by continuation of support based on deliverables and potential spin-off private community enterprises and cost savings.
Built	Building, Tools, and Technology	Material	Infrastructure such as roads, buildings, fences, sheds, and drainage systems.	Brick and mortar, such as pavilions, bicycle paths, water catchments for irrigation, benches, patios, bus stops, or toolsheds, that enhances education, recreation, and other community activities.

* The ethics and design principles are: Building, Tools, & Technology; Education & Culture; Health & Well-Being; Finances & Economics; Land Tenure & Community Governance; Land & Nature Stewardship. https://permacultureprinciples.com/flower

† The 8 forms are: Intellectual, Spiritual, Social, Material, Financial, Living, Cultural, Experiential. http://www.appleseedpermaculture.com/8-forms-of-capital

and they can accumulate or be depleted. For example, you may have heard of social capital. There is evidence that social capital provides a key entry point for improving community development in situations where extensive financial capital does not exist. Several other forms of capital also flow through, are exchanged in, and grow within communities.

Based on the Community Capitals Framework developed by Cornelia and Jan Flora, we provide in-depth descriptions of seven capitals in chapter 3: human, cultural, natural, economic, social, political (sometimes referred to as "institutional"), and built (sometimes referred to as "produced" or "manufactured").[4] The community capitals framework is not the only approach that uses capitals terminology and a range of constructs to make sense of asset stocks and flows in communities. Ethan Roland of Appleseed Permaculture has developed a similar set of eight investment capitals—social, material, living, intellectual, experiential, spiritual, cultural, and financial.[5] Many of Roland's capitals are similar to those in the Community Capitals Framework. Some differences exist, such as spiritual and experiential capitals, which we see as subcategories of human capital.

Toby Hemenway touched upon capital constructs in *The Permaculture City*.[6] His purpose was to demonstrate a way to evaluate holistic wealth, increase one's real value, and secure a livelihood that is more fulfilling than one based strictly on financial capital. There has also been a great deal of capitals-focused work in social permaculture through David Holmgren's symbolic flower, which defines seven domains of action that spiral upward to create a sustainable or regenerative culture.[7] Table 2.2 on page 33 is our interpretation of the relationships between the Community Capitals Framework, Ethan Roland's eight forms of capital, and David Holmgren's permaculture flower categories, along with a general summary of their association to community food forest initiatives. These three frameworks slightly differ, but the general pattern is the same: resilient communities are supported by diversified assets, and community food forests offer one way to invest broadly.

Thinking systemically about which community capitals to invest in is worth the time and effort. In the next chapter we will explain each capital in more detail to help you identify how they can build on each other and be exchanged, leveraged, tapped into, and created. This process will help when seeking to develop and manage regenerative and meaningful community food forests. The concept of stocks and flows provides a framework for creating a balance between idealized and practical community capital investments. In this way, community food forests can reach beyond their physical boundaries in support of civic well-being.

CHAPTER 3

Capital Investments in Community Assets

Thinking about project development and management across a diverse set of capitals can help a community food forest team identify partners,[1] understand community needs, and address citizen interests. Capitals provide guidelines for prioritizing change in a community.[2] They can be used to chart on-the-ground strategies to support diverse audiences[3] and mobilize assets to improve economies and environments.[4] When thinking about how to study community food forests and share results in a meaningful way, we were drawn to community capitals because they offer a framework for making sense of the complex relationships and broad benefits inherent in a community food forest. As far as we are aware, this is the first time they have been applied to urban food forestry, and we hope others will build upon this foundation and continue to refine the application. In these settings, we think community capitals bring clarity and organization to the fundamental but expansive processes of designing, establishing, monitoring, and evaluating a community food forest.

The set of seven community capital investments modeled by Flora and Flora provides one pathway for thinking about the spectrum of potential community development targets.[5] Their parsing of capital, from financial to cultural to built, can help tailor and balance diverse community food forest goals. By keeping a range of community development priorities in mind, leaders are better able to find specific opportunities and make refined progress. This by itself makes a difference but it also may tangentially affect other capital gains that are equally important in community development. Diversity and interconnectedness of community development goals avoids placing all of the project's eggs in one basket, so to speak. It positions leaders to think and act more broadly but all the while systematically. In this chapter, we first describe each of Flora and Flora's seven capitals in the context of community food forests and end by connecting these capitals to asset-based development, which is one way to align the priorities and strategies of a community food forest with the people and place it is designed to serve.

SOCIAL CAPITAL

Social capital is one of the most recognizable and discussed community capitals. In general, it is the trust, norms, cooperation, leadership, common vision, and networks that people share.[6] Social capital is complex and likely the most difficult type of capital to measure, yet it correlates well with the goals of most community food forests. Investments in social capital occur in two ways: via bonding or bridging. Familiarity and trust are the basis of both.[7] Bonding capital emerges out of the camaraderie of individuals coming together around similar goals and aspirations. For community food forests, they are usually people who already share expertise and values. In this case, some level of trust is typically already present. For example, one person may decide to join a community food forest project because another person invites them to do so, and working together on the project may further strengthen their connection.

Bridging happens when two people do not know each other or have yet to build trust, but are able to do so through a shared effort. In this case, it is the shared experience with the project itself that becomes the basis for building confidence and familiarity. In the community food forest setting, trust between and among participants is social capital we can see through acts such as borrowing tools, asking for a ride to a work party, swapping recipes, and making plans for a shared dinner.[8] You can invest and build assets via both bonding and bridging to strengthen existing and forge new interpersonal and professional connections.

Community gardens and similar community projects usually cycle through most activities and accomplishments on an annual basis. Community food forests provide benefits each year, too, but they will change over time because the forest is a dynamic, perennial socio-ecological system. People must adapt, shift, and respond as the forest grows. In the first years of a project, investments in social capital often come in the form of volunteer work parties and shared meals that correspond with site preparation, plant installation, and nurturing growth. After the initial excitement of site establishment, which usually brings people together with minimal effort, a shift in social capital investment will need to occur to positively affect other community assets. In these later phases, investments in social capital are matters of cooperation, persistence, and leadership. For example, scheduling regular maintenance parties, a variety of food production and environmental management classes, and other community events will keep people connected to the site and each other.

When thinking about social capital as a potential investment, it is important to consider the scope and size of a project and take stock of site conditions and community characteristics. For instance, if a civically engaged neighborhood with strong social ties plans a community food forest, assets of trust and common vision may not need attention. The project may simply fit into an existing community trajectory that is rich in social capital. Other communities may have a dispersed social network limited by cultural or political differences. In these cases, food forests are often proposed as a mechanism for increasing community assets such as cooperation and identity.

Social capital is often a prime focus in the early stages of community food forest projects. As a food forest grows and creates greater returns, this early attention to social capital can help kick-start investments in other assets. After project establishment, sites where bonding social capital was strong beforehand may need to shift focus to bridging social capital. Perhaps the focus should be trust and collaboration between partner organizations or addressing historical barriers to larger networking. If the social dynamics at a site seem to be functioning well, then investments in other forms of capital may be more appropriate.

HUMAN CAPITAL

Human capital is often a chief area of investment for community food forests. It constitutes the skills, experience, education, health, attitude, and other attributes possessed by people in a community.[9] Education is perhaps most highly studied in terms of how investment in human capital contributes to community well-being and economic development.[10] It is well known that there is generally a direct, positive correlation between investment in education (such as a college diploma) and increased opportunities for economic development (access to more and better-paying job opportunities). However, the education and experiences that come from involvement in a community food forest may not formally translate to financial gain. Building human capital through community food forests occurs when individuals increase their food literacy, participate in civic programs and local politics, eat healthier, improve their gardening skills, recreate more, and become better environmental stewards. By participating in the planning, implementation, and management of a project, community members learn skills in leadership, teamwork, and community organizing. Some volunteers may go on to organize people for another project that creates change in the community, participate in town hall meetings, or write grant proposals.

Assessing the range of human capital opportunities and needs in a community is an important part of planning for community food forest leaders. Without such an assessment, skill sets could go unnoticed because they were not given suitable opportunities to emerge. For instance, a sociology study using the Community Capitals Framework with Latino community gardeners in Iowa found that more than eight in ten grew up on farms.[11] Their skill represented a form of human capital that was tapped into for the benefit of all participants through coordinated and informal education and hands-on learning.[12] The merger of local production practices with those of immigrant constituents led to higher levels of efficiency and sophistication at the garden.

In the example above, making room for a different cultural approach to gardening increased cultural capital and strengthened social bonds between gardeners. This led to simultaneous investment in other capitals, growth in built capital (garden as green infrastructure and a constructed place), and better natural capital (environmental services and diverse vegetation within the garden). This demonstrates how investments in human capital may have a spiraling effect. In other words, a range of community attributes may continuously benefit and build on each other, leading to a greater overall positive impact. For example, when someone uses planting skills they learned at a community food forest to rehabilitate land elsewhere, they increase natural capital. Cultural capital, in the form of preference for food produced by sustainable production techniques or locally grown initiatives, may increase. As a result, farmers markets or other direct-to-consumer businesses may grow, which translates into financial capital for local producers.

Matching community skills with site needs and capacity can also be valuable in deciding what type of educational programming is needed. If the intended audience for an event is experienced growers, an environmental organization, or a permaculture group, no instruction may be needed. However, if one goal is to develop capable and environmentally engaged growers or instill a stronger sense of awareness about local food production, project leaders will likely need to design and implement specific educational programs to achieve those ends. For example, if an aim is social justice it may be important to have planning sessions that examine which foods have nutritional value but are cost prohibitive. Workshops can then be planned to show how to plant, care for, and harvest those species along with providing recipes and cooking demonstrations.

How can project leaders ensure they provide programs that benefit participants? One way is to focus on the value of educating the whole person, rather than only marketable skills.[13] Community food forest leaders tend to seek social change through holistic education using various learning methods. Designing different kinds of workshops and volunteer programs ensures there is something for everyone. Some people learn better by doing, while others learn by watching or applying something learned to a new situation. Some people respond well to a single teacher and direct instructions, while others do better in social learning circumstances where they grow and develop by interacting with others through discussion, observation, or mimicking actions. Understanding each of these educational methods can help plan events that create conditions for a broad range of participants, which can ripple onward to other capitals.

Workshops and volunteer days are also tangible metrics for grant reporting and quantifying impact. The number of opportunities to learn is a direct metric that can be counted and reported easily. Other metrics, such as the level of informal education and experiential learning that takes place when working with others during a volunteer day, are harder to measure and take more time. Creating simple surveys for "before" and "after" assessments of knowledge or offering a way to provide feedback about an event are a couple of ways to collect information that inform metrics regarding returns on investment in human capital.

NATURAL CAPITAL

Natural capital is the tangible and intangible aspects of natural resources and environmental amenities. Natural resources such as soil, water, plants, animals, and air have value for the life-sustaining services they provide. There also are environmental amenities and benefits that come from natural resources, such as prized viewpoints, construction material, clean air and water, and outdoor parks for recreation.

Food Literacy

A local food literacy workshop can introduce people to specific food-producing plants. Other topics might include difficulties in commercial production and transportation or which local stores carry sustainably sourced products. A workshop to increase local social literacy could follow the production pathways of specific produce in a community food forest, such as organic raspberries. Participants could pick a store nearby where they are sold and then critically examine who has access and buying power to purchase organic raspberries there. Map the supply chain along which those raspberries are grown, harvested, transported, packaged, and finally sold.

Ecological Function and Environmental Services

Ecological function is the natural environmental processes rendered through ecosystem relationships. When impaired, ecosystems lack the ability to fully function. Communities with impaired ecosystems generally fail to realize all the beneficial services that are possible in an intact ecosystem. Worse, highly impaired ecosystems turn services into liabilities. For example, air pollution can increase asthma rates. Food forests are designed to optimize ecological functions that provide food and environmental services, but if a site is impaired, it is important to select species that help mitigate problems before edible species are planted.

Community food forests invest in natural capital by increasing vegetation, improving soil, and providing food. The natural capital derived from a community food forest contributes to environmental assets that benefit urban places. These include reduction in air and noise pollution, enhanced biodiversity, microclimate regulation, heat-island mitigation, and storm water management. An impressive range of initiatives across the spectrum of community food forests demonstrate investments in natural capital, from increasing public access to green space to helping support technical water conservation projects.

Land is a necessary component for a community food forest. Unfortunately, projects are often given marginal sites that are undesirable for other forms of development or long left vacant, degraded, or covered with waste. In the built environment, vacancy and waste are important issues affecting a community's natural capital, and many community food forests seek to convert vacant lots or brownfields from liabilities to green spaces that provide a host of social and environmental services. While these are noble goals, there are risks: Degraded sites can contain harmful chemicals or trace metals from previous land use. There may also be water quality and management issues that require attention.

Although a site may seem perfect for a community food forest, it is always advisable to study the state of its resources and subsequent need for investment in natural capital. For example, studying the soil should always be a first step. Before a project moves forward in earnest, full disclosure of trace elements, chemicals, or other potential issues is necessary. Funds might be available for brownfield rehabilitation or water quality improvement. Natural capital activities can be safely invested in once site rehabilitation is complete. For example, at the Hazelwood Food Forest in Pittsburgh, leaders put in time rehabilitating soil by building organic layers on top of existing soil and covering the site with plants that quickly accumulate toxins. The plant matter was disposed of before planting of edible species began.

In the same way that capitals can build on each other in a positive direction, they can also spiral in a negative fashion. Natural capital can be severely affected by political capital, which can either facilitate access or create barriers. It can restrict potential uses of a site, delay implementation, and threaten or ensure long-term tenure. The impact of clashes between use of natural resources and politics are well-known, but less so are the spiraling effects on human, social, cultural, and financial capital. In more than one community food forest we visited, navigating the politics of ownership, access, and use of a potential site caused a reduction of people willing to commit time and effort to the project, as roadblock after roadblock delayed initiation.

Finally, investment in natural capital (and other forms of capital) can shape personal values and a culture of stewardship. Some people enjoy a sense of meaning from protecting natural resources or passing on an environmental legacy. Addressing ecosystem impairment can be an important part of community food forest planning and management. Programs or workshops can be created to explain how the site is contributing to land rehabilitation and the value of environmental services being repaired. Participants can be encouraged to use their new knowledge and skills (human capital) to improve their home property and other areas where they live to invest in the growth of natural capital and initiate trends of an upward spiral of positive capital outcomes.

CULTURAL CAPITAL

Cultural capital exists in the views and experiences of people about the world around them, shaped and defined over time through community socialization. Community in this sense includes religious, ethnic, place-based, or family groups that share values and norms. There are also intangible forms of culture that people learn by growing up and being educated in a particular society. Exposure to a certain way of life, of doing things in a particular fashion, and to dominant thought patterns and accepted realities shape how one interprets their life experience. More tangible aspects of cultural capital that people often easily notice and relate to are events and artifacts such as art, poetry, music, technology, architecture, recipes, festivals, and rituals. Whereas human capital investments are generally about improving an individual's quality of life, cultural capital investments are about making life more meaningful for people within a community.

Creativity is an important aspect of cultural capital, too. Including art—such as sculptures, murals, crafts, and music—in the food forest can shape community conversations and express values in a vibrant, revitalizing way and allow communication across language barriers. Communities that welcome diversity will attract more of it. This forms a regenerative cycle that broadens viewpoints and creativity for more successful solutions to community issues. Interestingly, several of the larger cities in the United States that rank high in terms of creativity, including San Francisco, Austin, Boston, and Seattle, also have community food forests.[14] It is important for a community food forest to be intentional about how it invests in creativity and diversity through projects, plans, events, and installations because its culture will shape how the public views the site and the groups involved.

What are some of the ways that community food forests can help invest in cultural capital? One way is by simply establishing demonstrations of plants and trees that grow well locally, produce food, and help a community buffer against changing environmental conditions. Another way is to plant species with local traditional uses, historical value, or folkloric medicinal properties to connect past uses with the present in support of collective community heritage.

The availability of nontimber goods can underlie a culture that values handmade products and self-care. People who forage usually do

> ## The Power of Repeating Celebrations
>
> Planning community celebrations around planting and harvesttime is a powerful investment in cultural and social capital. Annual events create renewed attention and appreciation for the community food forest as an important place and community resource. The Bloomington Community Orchard celebrates a cider festival every fall. The event is open to the public; visitors can try fresh cider and make their own with an old cider press. It introduces people to a traditional practice, adds value to apple harvesting, and is fun. The Hale-Y Community Garden Food Forest in Blacksburg, Virginia, dedicates time in early spring when participants come out to help stage the first loads of municipal compost, share hot cocoa or coffee, and build enthusiasm about the coming summer. It is an opportunity to reconnect after winter months, as well as work with and celebrate seasonal cycles. Any celebration or event that is repeated on a regular basis helps create tradition and strengthen bonds between participants.

so because of interest in a lifestyle more closely linked to nature. In other cases, it is tradition in their family or country of origin or is based on the need for food or naturopathic treatment. A demonstration of wild-crafting—gathering herbal, folkloric, edible, or ornamental plants (nontimber goods) from a forested area in a sustainable way—teaches people how to collect such goods thoughtfully and supports a culture of stewardship.

A community food forest that provides groups a place to celebrate important occasions, cultural heritage, festivals, and family events can strengthen and grow community by publicly displaying that such things are valued. Planning for these events reflects a community's shared vision and sense of meaning related to common traditions, beliefs, and community legacy. Events can even include demonstrations of locally fashioned horticultural techniques and tools. At the Hale-Y Community Garden Food Forest in Blacksburg, Virginia, participant ethnicity is rich and diverse. People are invited to make a traditional dish for once-a-month potluck dinners at the site using an ingredient found in the food forest or garden. During the meal, people explain the significance of the dish or other uses of the plant along with a short presentation on their country of origin and culture. While this is a specific way a community food forest can contribute to or shape cultural capital, it is important to note that much more broadly, projects are shifting how society experiences green space and considers food accessibility and abundance. And they are doing this simply by creating public spaces where produce and other goods can be freely harvested.

Investing in cultural capital in terms of shifting people's perspectives on land use, food abundance, alternative methods for production, and the possibilities of community collaboration and action can have major impacts on other capitals. However, cultural capital is often much slower to show results at the community level when compared to other individual and immediate investments, such as creating art or hosting a festival. When planning investments in cultural capital, short-term impacts should be designed so that they can contribute to the longer-term change community food forest leaders seek to create.

POLITICAL CAPITAL

Political capital broadly refers to power. The ability to influence the drafting and enforcement of rules and regulations, development projects, and issues affecting the community are a form of political capital. When it comes to community development, power is the ability to influence decision making.[15] It directly affects community food forests, urban agriculture policies, and the use of marginal urban areas.

When designing a community food forest, two forms of political capital project leaders should consider are instrumental capital and structural capital. Instrumental political capital includes the resources a project can use to influence policy.[16] It comes in the form of funding for particular initiatives. It is also the human resources, which often rally to mobilize and advocate for favorable perspectives on particular issues. Leaders should strategically seek out participants who are talented at organizing people for change and advocacy. Social media can play an influential role in community action, public opinion, and rapid information sharing. Someone with social media skills is an asset to any team; such a person can be called on to build a fund-raising campaign or publicize other community food forest needs.

Structural political capital refers to public perspective regarding the formal and informal aspects of a political system, such as laws, legally fixed procedures, or even shared meaning on issues. It influences the ability to access and accumulate instrumental capital, along with the ability to use it effectively. On a large scale, structural political capital affects community food forests and many other assets in a community. It also exists, and can be quite impactful, at the microlevel within a site. Laws and governing structures that impact a community food forest, whether formal or informal, actual or perceived, can limit potential access to resources. For example, some community members may have reservations about gathering at a site because they perceive barriers between themselves and site leaders or active volunteers, some of whom may not be from the area. What is more, development of a once empty space may imply something that is privatized and off-limits. Even when signs say otherwise, perceptions about access and exclusion may inhibit some community members from participating and thereby diminish local impact. For example, some community food forests are located in high-crime areas and are often established by outsiders with good intentions. It is not uncommon, however, for residents who live nearby but are not involved with these community food forests to experience some level of confusion or hesitation about whether they can use or volunteer at the site. For example, Cathie encountered a community member who was reluctant to use a food forest in her neighborhood. The woman passed the site daily and often wished she could harvest some of the herbs. Signs at the site noted that it was freely open to all community members, but even so, she still had reservations. She decided against going at night because of safety concerns. Moreover, the only people she had seen tending the site were not from the neighborhood, and she did not feel comfortable harvesting without their permission. This is a case where, despite multiple signs encouraging public access and use, the *perceived* rules governing a site can restrict bridging across communities of place and interest, which can significantly inhibit broader inclusion.

Community food forests facilitate direct participation in decision making by offering people the opportunity to become involved in governance of the site. People can learn how to interact with local officials to obtain permissions and other needs. And people can learn analytical skills such as how to question structural variables, test assumptions, and create new structural conditions.

Thinking strategically about the forms of political capital that affect administrative cooperation and support for a community food forest project is crucial. Alienation in a political sense can be catastrophic. Project leaders should revisit political investments and associated assets throughout the development of a project to stay ahead of the politics affecting a site. This process primarily focuses on anticipating levels of community in adjacent neighborhoods to ensure people do not feel alienated.

Community empowerment and local support can increase as the presence of community food forest stakeholders and their relationship changes with local boards, county, state, federal, tribal, or regional governments.[17] A community food forest will always itself be a source of politics regardless of intentions. Thus, project leaders should be judicious in their politics, but stay true to their vision.

BUILT CAPITAL

Unlike other capitals, built or physical capital is mostly fixed and immobile. It refers to physical features such as buildings, roads, park benches, and water fountains. An investment in built capital generally stays within the community. In the case of community food forests, built capital

> ### Signs, Signs, Everywhere Signs
> We observed that one of the most important types of built capital investments community food forests have made are interpretative and informational signs and labels that highlight plants and other interesting features. They promote education and community acceptance and build networks of support and interest. Signs share key information about the project, place, vision, and concept. Installing them is also an important and effective way to provide guidance on etiquette and harvesting and to warn against potential liabilities such as food allergies or the presence of bees and other stinging insects.

normally refers to the infrastructure that supports use and maintenance, such as walkways, picnic areas, fences, sheds, or structures for meetings. Also unlike other capitals, built capital is usually highly tangible. It is easy to quantify investments (easily priced most of the time), and return on the outlays can be formally analyzed. Similarly, built capital, like other capitals, may degrade and depreciate over time. Sometimes it is vandalized or stolen. Or it may simply wear out and need to be recycled or replaced. Because of this, both short- and long-term investments in built capital are required. Community food forest initiatives should include plans to establish built capital on-site, whether donated, purchased, or even grown on-site using a "seed to structure" approach with specific plant and tree species. These plans include a long-term vision regarding how to maintain and expand the role of built asset investments.

Investments in built capital can also be a catalyst for development of other types of capital. For example, investing in a gathering space that has tables, seating, and cover or in an area with worktables and a sink to process harvested goods can lead to development of social, cultural, and human capital as people share recipes, celebrate traditions, learn how to prepare food, or play music and tell stories. Many community food forests also include a shed to house tools. This not only makes maintenance tasks easier, but provides the equipment for learning new skills that contribute to human capital. Other community food forests have created a place to hold small concerts or events. This can lead to financial capital by selling tickets for fundraisers or community parties at the site or by renting out the space.

FINANCIAL CAPITAL

Financial matters are some of the most important and consequential concerns for community food forest projects. In most cases, grant funding is a critical early step; it can help tremendously with planning and go a long way toward covering the costs of project start-up. In general, though, funding is important across multiple implementation phases to secure equipment, acquire planting material, and construct benches and sheds.

Showing access to other forms of capital, such as the ability to garner community support and labor, often increases the possibility for obtaining financial support. Funding institutions typically prefer projects that demonstrate community impact and leverage human, social, and political capital as in-kind contributions, in the way of gifted or low-cost land, donated planting stock and equipment, and volunteer labor. Participation in a community food forest can generate financial capital in other ways, too. As people learn production skills on site, they may take those skills home. They may reproduce a food forest on their own property and increase family savings by augmenting food needs through perennial gardening. They may even develop a small business based on products from their personal food forest.

Determining which types of financial capital to invest in can be complicated. People's personal philosophies, experiences, and skills related to money management vary widely. It is important to keep in mind that useful and impactful investments usually include a balance between realistic monetary and human investments. Community food forests need money, but their goal is to be freely open to the public, not to act as money-making ventures.

Three forms of financial capital are relevant to investment strategies for community food forests: equity, debt, and venture. Equity capital comprises direct and permanent investments that shape a project and fuel work using familiar and immediate funding sources. The money needed to purchase planting stock, protective tree tubes, fencing, lumber, gravel, sand, and so on represents a direct investment in site permanence. Debt capital is a short-term credit option that helps with project investment both on- and off-site. These types of debt-based capital come with complications of loans and interest, so the risk is greater. Venture capital is an investment in a nonguaranteed enterprise. In some cases, there may be environmental venture investors that account for return on investments in ways that include forms of capital beyond finance, such as environmental remediation, but this is rare.

Transforming Capitals

One of many examples of entrepreneurs making a living from skills learned at a community food forest is Lincoln Smith, owner of Forested, LLC, in Bowie, Maryland. Smith helped design and install multiple public food forests in eastern Maryland as part of his company's work. Additionally, he manages a ten-acre food forest near his home. There he holds workshops, runs a CSA, and raises funds through an annual harvest dinner featuring local chefs. Harnessing his landscape architecture design skills and agroecological knowledge, he has built a business that supports and engages people in food forests throughout the area. Smith's human capital investments have also transformed the food forest experience into political change. His work on edible landscaping in other towns resulted in building code analyses in those towns with an eye on improving opportunities to grow perennial food-producing species in unused space.

Community food forests usually depend most heavily, if not entirely, on equity capital, either through philanthropy, grants, or investments. Some community food forest projects we visited were fortunate enough to have substantial upfront contributions of equity capital. However, obtaining financial capital investments at the outset of a community food forest project may be very difficult. It will require creative thinking about sources of funds and operating on a shoestring budget. In-kind contributions of materials can help when money is short. Efforts can pay off over time, as the project and site take shape. Establishing a reliable track record will ultimately garner interest from other stakeholders.

For most community food forest projects, it is wisest to seek direct contributions and potential low-interest loans, if needed, to carry out necessary investments. However, receiving grants or large donations early on can be a mixed blessing since they typically come with hard deadlines, time-consuming reporting requirements, and the need for creating and sticking to a schedule of events. If plans are not yet well defined or if the network of social capital is still small, the work of establishing the site may fall on the shoulders of the leadership team alone. There are cases when the source of financial capital has had heavy influence on the design, establishment, and scale of a food forest. Our advice is to be strategic in deciding when it is best to seek a grant and be ready with plans to implement one within a long-term vision.

PUTTING THE CAPITALS TO WORK

Through participatory processes like community food forest workdays, people build human capital through interacting and sharing knowledge. Working together for a common cause underwrites human, social, and cultural capitals. The food forest itself is a springboard for natural (such as plants, habitat, and soil) and built (such as creating a gathering space or park) capital. In short, our observations of the complex interactions between capitals in a community system demonstrate that a community food forest's success is not linear. Nor should you think of capital investments in a linear or compartmentalized fashion. They do not follow a prescribed sequential pattern occurring all at once or in the same way at every site. And they do not operate in a vacuum or silo.

Asset-based development is one way to identify and mobilize existing assets in a community, build strong relationships between stakeholders around these assets, and involve community members in making decisions that enhance and grow them. Speaking of benefits in terms of assets can be a powerful way of describing a project's foundation and highlighting and leveraging its strengths. We have found that much of what community food forest leaders shared about project outcomes can be defined as assets in a clear and distinct way. Asset mapping connects the dots between community strengths and pinpoints shortcomings and needs. When used in conjunction with community capitals as a guiding framework, asset maps help community food forest teams spell out and leverage tangible elements such as open land or a community need, as well as intangible factors such as skill sets that stakeholders offer. Community capital gaps in an asset map also define pathways for improvement.

Thinking in terms of capital beyond a financial context opens possibilities for a more holistic understanding of community investments through a public food forest project. Potential dividends likewise expand as is evidenced by the case studies in this book. In that regard, you may find it helpful to jot down examples of the capitals and their relationship to each other as you read through their stories. We would also encourage considering how you might translate your notes into asset-based messages that could be used to win over more advocates and strengthen a community food project's marketability and impact. The following real-life examples demonstrate the diverse range of patterns that community capitals follow in the planning and execution of community food forest projects we visited and provide a few short examples that you can consider before taking in longer case studies later in the book.

Mercy Edible Park, Philadelphia. City code revisions that accommodate new land use strategies increase the multifunctionality of a valued commodity in urban places. Many advocates for such revisions also promote policy that supports vacant lot rehabilitation and city revitalization with green infrastructure projects. Investments in political capital along these lines are paying off in Philadelphia. The founder of Mercy Edible Park, Robyn Mello, went on to become the director of education and outreach at an organization focused on planting and caring for fruit throughout the city. Robyn's advocacy and leadership across key political investments circles back to the community food forest by spiraling out through investments in human and natural capital through her work.

18th and Rhode Island Permaculture Garden, San Francisco. The leaders of this project situated on borrowed property are working to change policies so that owners of city property are able to receive tax incentives when they allow agriculture on unused properties. The community food forest is on property freely leased by a local doctor interested in public health. To ensure long-term success, project leaders have been building support for California's first state agricultural incentive zone. This will help make developing unused property for food production easier. The idea is that property owners receive credit for not developing the property; neighborhoods benefit from increased green space, which typically increases property values and well-being; and community members have

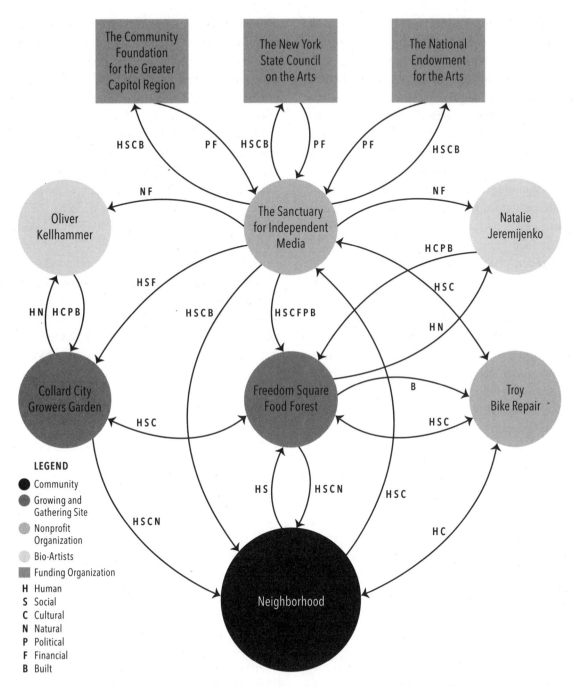

FIGURE 3.1. This is a systems map of stakeholders and the capitals that flow between them. Not all stakeholders and capital exchanges that make up the whole system supporting the Freedom Square Food Forest are represented in this sample. The Sanctuary receives funding from multiple cultural and arts institutions, which they transform into support for their gardens, food forest, and neighborhood by hiring environmental artists to be visiting residents and base their projects in the community resources.

a place to recreate, relax, and harvest food. On top of this, the community food forest serves as a base for a permaculture group that uses it for demonstrations and workshops as part of their business model (and in return, the group takes care of the food forest). All told, this example represents a case of political investment improving financial capital. This, in turn, improves opportunities for urban agriculture.

Bloomington Community Orchard, Bloomington, Indiana. Investment and success in social networking are also common community food forest outcomes. In Bloomington, for instance, the community-based organization Bloomington Community Orchard helped create a remarkable partnership between local government, academic institutions, and community members in support of the town's community food forest. They also leveraged cultural and social capital at Indiana University to reach alumni beyond the local population. They built support, in the form of online voting, in a national orchard project proposal competition whose winner would receive an implementation grant. As the orchard developed and interest in perennial plants on public land grew, it began to transform into a food forest with the addition of more and more plants. Local artists incorporated their projects, too. Today, events such as yoga sessions and annual fall harvests are held at the site, which contribute to human and community health in the town.

Rahma Edible Forest Snack Garden, Syracuse, New York. In Syracuse, Frank Cetera, founder of a community food forest and permaculture/food-not-lawns organization, increasingly works with other initiatives and groups leading efforts to reduce poverty and improve living standards. Social capital created at the Rahma food forest seeks to increase engagement among community members for other projects that are not related to the food forest but require similar investments of community capital. Learning to organize people and advocate for meeting the needs of underserved community members are forms of human capital. The strategy Frank has been trialing is a stewardship committee made up of members representing the organization, the landowner, the neighbors, and a broader community development group.

Freedom Square Food Forest, Troy, New York. The Sanctuary for Independent Media, a multipurpose nonprofit, created a community food forest as just one of their community development projects. The Sanctuary receives grants from various organizations to support programs in which they creatively integrate environmental, educational, cultural, social and community health aims. They collaborate with another local organization, Troy Bike Repair, to accomplish some of this. Troy Bike Repair offers programs that teach local youth how to fix bicycles, with the reward of earning one.

Parents were concerned about child safety in the community because of crime, so the Sanctuary added a network of trails throughout its gardens and food forest that children can use for riding the bicycles they have earned. These trails provide a place where children and teens can play safely, while interacting with annual and perennial food systems. The program has reduced the number of bikes stolen in the neighborhood. Children and teens have become interested in food production and the environment. They have also learned how to express themselves using various forms of media.

Richmond Edible Forest Project, Richmond, California. The nonprofit organization Urban Tilth established the Richmond Edible Forest Project along a neglected urban greenway. The project created places for the community to congregate. Urban Tilth uses the food forest as an outdoor classroom where they offer environmental education and food production training. Local youth participate in summer programs where they learn how to use the principles of ecology to design food-producing systems modeled after a forest. Participants also go camping at a state forest. There, they expand their learning to an actual forest ecosystem. They experience nature outside of an urban setting, learn camping skills, and strengthen social bonds. Then they bring this information back to their neighborhood. They agree to be responsible for tending the food forest, recording data on health and productivity, and coordinating site irrigation.

Tree Streets Food Forest, Johnson City, Tennessee. One organization spearheading initiatives to use food forests as a genetic seed repository helped establish Johnson City's community food forest. The project aims to provide locally adapted species for the community and nursery businesses. The food forest volunteers map and observe low-maintenance, easy-to-grow edible perennial varieties around the city. Volunteers harvest seeds from these species for planting in the food forest. There they can be propagated and sold to people who want to grow locally sourced edible perennials on their own property. The organization is also training interested individuals on how they can begin their own tree nursery businesses. Some of the seeds collected from the food forest's local genetics observation program will be part of their supply. In this way, the food forest is contributing to economic development as well as to the natural capital of Johnson City.

As we described earlier, a community food forest is often just one of several food-producing initiatives nested within a larger community development system. It is wise to sketch out the overarching system and the relationships of other components to your community food forest, whether your project is in the very beginning stages or already underway. It is never too late for this type of analysis and reflection, because it can help reveal ways to develop and nurture your project even in the absence of traditional forms of funding and development. More broadly, community food forests and related projects are part of a transition in our culture. We see a change from individualistic lifestyles and mind-sets to ones that create a sense of local abundance and shared opportunity.

PART 2
Meaningful Planning

CHAPTER 4

Planning Fundamentals

This chapter lays out our findings about how best to get started with your community food forest project. We begin with a set of general project management phases and describe how those phases specifically relate to community food forests. We also delve into adaptive management, a cyclical process that constantly assesses project status and adjusts management to changing conditions and new information. It allows for a more flexible approach tailored to a project site while still following a basic set of management criteria, which we describe below in terms of phases. Community food forests are long-term initiatives, and an adaptive approach is part of a prudent plan that stands the best chance of sustaining momentum beyond early stakeholder excitement and volunteer energy.

One critical aspect of a project is whether it stands alone or is part of a larger community development program. The distinction is important because it impacts funding, access to support networks, and limitations or opportunities for project development. There are also differences in how a stand-alone project is managed as compared to a project within a larger program. Projects have timelines with a set beginning and end. These are agreed-upon deliverables or results. Programs tend to be ongoing, involve continuous process improvement, and may include multiple projects. We cover the implications of this distinction more fully in "Independent Versus Nested Projects," beginning on page 60.

Doing your best to anticipate a few fundamental management pitfalls can help avoid criticism of your community food forest and steer clear of some typical potential barriers to long-term establishment. At the sites we visited, project leaders confronted concerns such as safety, long-term maintenance, policy limitations, and difficulties adhering to town codes. We recommend checking into whether a project similar to a community food forest has been carried out in your town or somewhere nearby. If so, that project team may have already encountered and overcome hurdles or paved new pathways in relation to local policies. Connecting with others who have been through the process is a great first step.

PROJECT MANAGEMENT PHASES

Our experience studying and leading community food forests taught us that starting with the basics and seeking to understand fundamental management phases leads to effective planning. This is true whether the food forest is an independent project or part of a larger community initiative. Thinking about phases helps leaders identify and plan for when, where, and how to direct precious resources instead of trying to do everything at once. We will provide a general overview of the following five main phases associated with project management and relate them to community food forests:

- Initiation
- Planning
- Establishment
- Monitoring and maintenance
- Closure

The phases are listed in the order they occur when applied to projects that have a linear timeline. However, we found that these phases often did not occur in a linear fashion for community food forest projects. For example, one section of a site might be in the establishment phase, while another previously established section is in the monitoring and maintenance phase, while yet another area is being planned for a future planting. These phases do not only apply to the installation of plants, either. Volunteer education could be in the planning process while another is currently being implemented. At the same time, the steps to close down and document a recent workshop might be taking place while future visioning is happening for the next. Soliciting feedback from participants and deciding whether or not to offer the same or a similar workshop in the future is part of closure. The information gathered provides direction for planning the next volunteer education phase.

Initiation

Initiation is the origination and early evolution of an idea. It starts when the idea is born and gets underway when the thought is shared with others to test the potential. With positive feedback, excitement begins to build. This is typically the phase when people reach out to those they know could be interested in the idea to gain support and encouragement. This is an exhilarating time for a community food forest project. Brainstorming is a great and exciting way to generate ideas and start working on details such as what species will be desirable. It is tempting to start on a detailed food forest design, but typically a full design is a product of the planning phase due to the importance of first acquiring land and seeking community input. While visiting food forests across the country, it became evident that it is next to impossible to predict how many people may be involved when initiating a community food forest. There may be scores of people involved or just a handful and it is best to prepare for both.

Identifying your initial objectives is helpful to avoid a sense of being immediately overwhelmed by the overall scope of what needs to be accomplished. Let us break down some of the initial steps of initiation and consider how they relate to community food forests.

Property. One of the first details to consider is space. Is property easily available? If not, what will it take to locate an appropriate place and negotiate a lease or tenure? At this stage, it may be helpful to seek advice from professionals such as those mentioned in chapter 9. For example, an urban forester could specify green spaces where increasing canopy coverage is possible, or an urban planner may know of suitable land where future development is unlikely. The owner of a vacant lot may be able to be contacted by checking a city's vacant property registry. Finding the best location and securing permission can be a lengthy process. It is best to start early as the space selected will influence the participants involved, the planning process, and dynamics with the broader community.

Community context. Initiation is also the ideal phase in which to think about historical and contemporary issues in a community and the role a food forest can play in adequately and fairly addressing both. Is there a historical context that will provide cultural meaning to the design? Is there a history of neighborhood division to be bridged through careful planning? Or is there strong unity in a neighborhood that will support the project and make selecting a location easy? These types of questions are critical to early success; in chapter 11, we offer more of these important questions and discuss why they are important. Another detail to reflect on early is whether enough people have the time, energy, and interest to pursue a community food forest and take it to the planning phase.

Mission and target audience. While brainstorming about your community food forest, an important early step is to ask yourself this question: Why this project instead of another? Critical reflection (discussed in chapter 8) is helpful for this type of discussion. There are multiple ways to increase food security in a town, build community, or act on behalf of social justice. Because of this, it is worthwhile to explore questions such as: Who will the community food forest serve? Why is a community food forest the most appropriate solution for the people it will serve? Systems thinking, community capitals, and critical reflection are helpful in these brainstorming sessions. After identifying why a community food forest is the most appropriate project, leaders can go on to identify the scope.

Scope. Scope sets limits around what the project can and cannot address. It focuses a project by recognizing it cannot be the solution to every problem. Important when defining scope is to revisit project vision and objectives. The vision will go through multiple stages of development and the first iteration is one that gains peer approval to pursue. During the pursuit, there is a need for collecting feedback and opinions from

Participant Turnover

Community food forest leaders must think about planning for human succession and community context. There is a broad spectrum of planning needed for these projects to develop a solid proposal. Sharing the facets of such planning between multiple people spreads, and thus sustains, the project knowledge but can be overwhelming when leaders are simultaneously managing multiple phases of development and coordinating everyone. If knowledge of the inner workings rests solely with one person, however, the community food forest will not thrive. Project leader turnover is inevitable, and ensuring that institutional information is archived and shared is a key step in the success of any project.

Volunteer turnover is par for the course, too. People who are in transitional periods of life may participate once or twice and then never return. A constantly changing pool of volunteers can be difficult for an organization or project. Planning ahead for specific dates to host other established volunteer groups, such as Master Gardeners, AmeriCorps, Student Conservation Association, or Boy Scouts, is one way to ensure predictable pools of caretakers.

others about the vision. Future iterations should then be collectively developed and shared so that everyone sees their part reflected and is willing to work together toward making the vision a reality.

Setting the scope of a project also makes it easier to identify who should be involved and define their roles and responsibilities. We have included a set of archetypal roles for your reference in "The Core Group," on page 174; this positional lineup is a helpful starting point and can be used to think about who can fill critical roles from the outset. Once you have a better idea of who will be involved, you and fellow leaders can form an organizational structure or decide on a decision-making process. Because decision making often needs to begin almost from project inception, it is better to agree upon a decision-making system well before the planning phase.

Feasibility. Lastly, as part of feasibility discussions, leaders need to consider risks, assumptions about a project's aims, and other issues that potentially will affect the community food forest. Being realistic about these components allows the group to plan for how to avoid or deal with them if encountered during the project. For example, with any community food forest, there is a risk that the organization or group of people who initiate the project will either dissipate or move on. In chapter 7, we describe how this happened at the Dr. George Washington Carver Edible Park in Asheville when the founders of the organization moved from the area. Because Asheville agreed to share responsibility early in the process, the park was able to weather the transition to a new caretaker organization under the oversight of the municipality. This is an example of how site continuity can be preserved through planning with a public partner who agrees to share some of the risk.

Planning

Sometimes it is difficult to determine when initiation stops and planning begins. Usually, there is a transition period when they are blended. Initiation might need to occur in a couple of months or it might take several years until all necessary approvals are in place and in-depth planning can begin. During this time, planning on specific aspects may be underway. The planning phase grounds the details developed during the brainstorming and stakeholder analysis in the initiation phase. Defining project schedules, agreements, and communication strategies is part of planning. Leaders will work with other stakeholders to develop a realistic timeline of events and dates for resources, equipment, material, labor, and other necessities. In certain regions, the timing for acquiring plant material, transplanting, or direct seeding is determined by climate, which makes it easy to develop a timeline. On-the-ground food forest layout and plant establishment generally need to occur in phases spanning several growing seasons.

Planning sets up the next phase of action and community food forest establishment. In chapter 8 we discuss agroecological concepts that provide useful guidance on ideas, design principles, and practices to incorporate into a plan. We present the story of the Wetherby Food Forest, in Iowa City, which demonstrates the importance of starting small and expanding over time as a site proves sustainable. Planning also includes thinking ahead about how to fill gaps in planted areas that may occur due to plant mortality. Keeping future needs in mind such as filling in gaps helps plan for managing things such as when and how much funding to set aside or how and when to organize a community drive to accept plant or money donations.

Finding Advisors

Local food system projects are good places to find people who know about policies, organizations, or social networks that a community food forest project can utilize for further advice or volunteer support. Individuals, organizations, or websites associated with the following are good places to focus your research:

- Local food policy councils
- Cooperative extension offices
- Community gardens (try www.acga.org)
- Community-supported agriculture
- Farmers markets
- Social justice advocacy groups

There are also offices in local governments that could be helpful to contact for advice, including parks and recreation, urban forestry, sustainability (sometimes referred to as planning and sustainability), public works, arts commission, and fish and wildlife. Refer to chapter 9 for further information on allies to look for when kicking off the planning process.

In addition to site design, it is important to plan for how the site will effect change or serve a community population. During the initiation phase, the relevance and purpose of your community food forest is determined. Developing a thorough strategy for how that change will take place through initiatives tied to the community food forest is central to the planning phase. To help develop this strategy, we present the theory of change and ways to measure whether the project is effective in the next chapter.

Financial planning is critical and may require thinking a year or more in advance. For example, it is important to plan ahead regarding deadlines related to grant cycles and timing requests to municipal allies in order to meet the requirements of governmental fiscal calendars. Sources of funding also help determine what constitutes success along with metrics on how to measure progress toward it. Stakeholders often have different criteria for success. Being realistic about what is possible with the funding available and working out an agreement on what an acceptable level of success is for specific criteria will help everyone land on the same page about project expectations.

A communication plan often includes multiple levels. The core team will need to decide on methods of communication among themselves and also how the team will engage with the public or others. For example, the Mesa Harmony Garden in Santa Barbara, California, sends weekly emails to their listserv to update everyone on work that needs to be done, important events, and general news about what is happening at the garden or what is ready for harvest. Other sites, like Beacon Food Forest in Seattle, use social media sites to post information, events, and photos to entice continued engagement.

Establishment

In typical project management, this phase is referred to as the execution phase, but execution at a community food forest is better described as establishing the physical site and conducting projects that create conditions for vibrant community experiences. Overall, this is an action-oriented phase when plans are put into motion. Following through on a communication plan is one of the biggest tasks in this phase. Within the core group, communication should be constant. Updates on progress, setbacks, and general feedback needs to occur regularly. Communication with volunteers outside of the core group is also critical in this phase because volunteers are most often the labor and energy source that take community food forests from a design to an established physical project.

Monitoring is largely represented in the next phase, but it begins during establishment to assure quality of work and project status and, even more important, to celebrate achievements! Feedback loops begin during this phase as information is collected on what works and what does not. Checkpoints should be scheduled often, in order to pause, assess, and then inform project partners and funders of how the project is unfolding vis-à-vis the plan. Everyone involved in a project or program has expectations for progress, so managing to keep them realistic and agreeable is crucial. This is particularly the case during establishment, which transforms your community food forest concept into reality.

Monitoring and Maintenance

Monitoring and maintenance entails ongoing upkeep of a community food forest and periodic evaluation of its progress. The work of monitoring the social system can be even more important than maintaining the plantings. Expect an ongoing need to monitor how volunteer workdays are scheduled, who will lead them, and whether volunteers need training in pruning, coppicing, weeding, or other techniques. Learning more about volunteers and assessing needs can help stakeholders decide whether or not workshops in gardening, design, construction, vegetation maintenance, and so on are needed.

If maintenance goals are not being met, adaptation becomes necessary. In terms of physical maintenance of the site, protocol should be in place for how underperforming, dead, or dying plants will be replaced or rehabilitated and new plants added. It is also a good idea to set guidelines for maintenance and monitoring of infrastructure such as trails, benches, signs, and picnic tables.

Closure

Closure in a community food forest project can signify success, such as a desired level of community participation or a particular volume of fruiting. Closure can also be the end of a particular phase of the project, such as initial installation, or a predetermined point in time where project goals are revisited. We noted during our research that the idea of closure did not reflect the goal for community food forests to have long-term viability. This made discussing closure or pinpointing when it was happening confusing and difficult. It was easier to recognize times of closure in relation to discussions beginning about future visions for the next phases of growth, which is reflected throughout the case study chapters.

Documenting the project can be of great value during closure and other phases. It is helpful to decide in advance what archival procedures you will follow to record accomplishments and provide summaries of lessons learned. For example, the development of the Dr. George Washington Carver Edible Park was chronicled in a series of newsletters, which turned out to be very helpful because they created a public archive.

Community members must also decide on mechanisms for measuring expectations and satisfaction as a way to make real-time project improvements. Closure at certain junctures may only be fleeting. The whole community involved with a food forest, not just the project leaders, must be ready and willing to adapt to unexpected events such as volunteer vacancies, new policies, and natural disturbances like windstorms or human damage from vandals or plain wear and tear. Instead of planning for turnover or the unpredictable as project leaders do, closure, whenever it may occur, reflects the frame of mind among all involved that succession and change is natural in a social system and adaptation is a wise policy.

Adaptive Management

Adaptive management is a strategy that applies part and parcel to community food forests, but is especially applicable when it comes to monitoring and maintenance. This term, often associated with managing natural resources, describes a systematic and iterative process of making decisions based on evaluation and feedback. In the case of a community food forest, adaptive management relies on consistent monitoring to determine whether the intended ecological and social functions or benefits are being realized.

At a basic level, adaptive management is a cycle where ongoing observations are collected and information on metrics is consistently gathered. Then they are compared against the project timeline for completing objectives. As a result, decisions are made about redefining problems, establishing goals, and taking action. When a food forest is not growing as expected, thinking adaptively will help identify a range of potential solutions that will create the right conditions for desired outcomes to emerge. If the observed problem is slow growth, maybe light levels need to be adjusted or nutrient availability changed. Perhaps competition

Planning Feedback Loops for Adaptive Management

Systems thinking can help support adaptive management. Creating a mechanism or instrument for collecting feedback provides information that may influence decisions. What does this mean in the context of how to take action during planning? A mechanism or instrument can be a verbal or written survey administered at set intervals (like six or twelve months) after engagement with the food forest.

Say the goal of a program is to have participants replicate what they learn on their own property. A survey could include simple questions to assess whether any practices have been used at home, and if so, which ones. If not, why not, and was there any intention to implement practices there? Did use of practices at home relate to whether the volunteer is a renter versus a property owner? Feedback from the surveys can be used to analyze progress toward goals and adjust educational programming and available support. Or feedback may point to a need to adjust the original objectives of the community food forest. It is beyond the scope of this book to cover how to develop effective questions for specific feedback, but additional resources are available at www.communityfoodforests.com.

between two closely located plants is too strong for them both to survive. Considering various solutions and how a change in one will affect the system as a whole is all part of adaptive management. Ongoing monitoring is needed because it provides ample information from which to make informed decisions.

Adaptive management also applies to the community food forest's social system. For example, one goal of a food forest might be to encourage people to implement practices learned during volunteer workdays at the site on their property or near their home. In order to evaluate whether this goal is being met, metrics should be identified during the planning phase with decision points to check progress. If progress levels are low in certain areas, it is an indication that the current strategy is not working and needs to be adapted or that unexpected challenges need to be addressed. This is an extremely brief introduction to the strategy. The community food forest website (www.community foodforests.com) lists resources that offer more information on using adaptive management to improve results.

INDEPENDENT VERSUS NESTED PROJECTS

When developing strategies for adaptive management, it is important to determine whether your project is better described as an independent, stand-alone project or one that is nested in or likely to be tied to larger programmatic or strategic community initiatives. An independent project has a solitary endpoint aim that is usually specific to the immediate food forest area and project partners. There is a level of autonomy and internal decision making about when objectives have been met. Independent sites may offer ongoing programs for community education or exist as a unique operation or site available for public use and interpretation.

Community food forests that are created as part of a larger initiative, on the other hand, may be subject to modification until a larger community impact is realized. There may be many periods of project fluctuation and adaptation because the endpoint includes much more than the project itself. Much of this flows from its integrated nature, which can lead to a greater number of voices speaking out about what can and should happen. The need to adjust course may be related less to a particular impact and more to the management of community input from parallel civic projects and citizens more broadly.

Independent: The Roger Williams Park Edible Forest Garden

When community food forests are established by an individual or small group as stand-alone projects, it is important to think about ways to build consistent partnerships. Maintaining strong volunteer participation will require multiple internal projects at times until all components of the food forest are established. The focus is on creating and defining a place that incorporates internal aims for social and environmental change.

The Roger Williams Park Edible Forest Garden in Rhode Island is an instructive example of the development of an independent project. While interning at the Roger Williams Botanical Garden, a University of Rhode Island student proposed the installation of a food forest on the grounds of the botanical garden for a senior honors project. Primary users were first envisioned to be Master Gardener volunteers from the University of Rhode Island, and the

FIGURE 4.1. At the front entrance of the Roger Williams Park Edible Forest Garden, the plantings create a transition zone between the existing trees in the background and the grassy area in front. Tall elderberry plants create the overstory while blueberries, raspberries, yarrow, and other plants create lower growing zones. A group of Master Gardeners plans workshops and activities that expand this site year by year.

site was to serve as an education and demonstration resource for these individuals. Once trained, the Master Gardeners in turn would work with community members to steward and harvest from the food forest.

The community food forest was a new concept to the Master Gardeners. The plan was to focus their training on establishment strategies and maintenance techniques. The student received a grant from the university's outreach center to attend a permaculture design course in return for teaching introductory permaculture to Master Gardeners and other community members. After becoming familiar with some of the fundamentals of permaculture, the group formed a collaborative design team including Master Gardeners, community members, university outreach staff, and founders of a local nonprofit and permaculture farm, Revive the Roots.

Project partners first conducted a site assessment of a property near a newly designed community garden within the grounds of the botanical garden. The design team then selected a location between the community garden and a large pond. The vision was to add biodiversity to the existing but scarcely forested riparian buffer along the edge of the pond. The design would catch storm water coming off the community garden and add layers of production through use of food forest species. The team also tied in a historic pathway following the pond's edge. Species selection was limited

FIGURE 4.2. This herb spiral provides a gradient of moisture conditions that different species require. The installation serves as an educational tool for volunteers at the Roger Williams food forest and keeps them interested in the site.

to natives, including market-viable plants with historical value such as chestnuts, pawpaws, and New Jersey or red root tea, a shrubby native used as a black-tea alternative during the time of the Boston Tea Party.[1]

Everyone involved in the Roger Williams Park project helped design a site plan based on a thirty-year strategy that would first produce herbal medicine, tubers, leafy greens, berries, and some fruit, with nuts, mulch, fuel, and other types of fruit and berries coming in later years.[2] They selected species to design food forest patches installed over four phases to increase cover beginning in 2012. Results of the first planting were evident two years on and additional trees were then planted. The site evolved further in later years by incorporating shiitake mushroom production, coppice, and hügelkultur (no-dig raised beds) and is continuously used as a living laboratory and educational space, spearheaded by Master Gardeners.

Nested: Freedom Square Food Forest

Community food forest projects can be part of larger community initiatives such as neighborhood revitalization or urban agriculture and business-based food security and nutrition projects. In these cases, the goal is to ensure that the community food forest serves a defined role relative to other associated projects and bigger picture goals. For example, the Sanctuary for Independent Media in Troy, New York, leads a

PLANNING FUNDAMENTALS

FIGURE 4.3. Another Master Gardener project at the site includes bee boxes for pollinators nailed to a telephone pole.

The community food forest project in Troy is part of a larger revitalization strategy envisioned by the Sanctuary and was created in concert with other projects such as a community garden, bioremediation site, and permaculture demonstration area. It was one part of a larger project to develop a place that neighbors can use every day or schedule for special events and community gatherings. In 2014, the food forest component consisted of edible perennial landscaping along the edges of a transformed vacant lot that is now called Freedom Square. Fruit trees were planted along two sides of the lot bordering roads, with a diverse set of understory species planted underneath the trees, all enclosed with protective fences. The vision was to create an outer food and forage zone surrounding a recreational area for community use. By adding a useful perennial border for many of the related community initiatives in Freedom Square, the food forest complements the Sanctuary's broad efforts to transform community and place.

AVOIDING BASIC PITFALLS

Thorough planning requires figuring out how to achieve all project goals and objectives as well as how to avoid or deal with potential pitfalls before they become obstacles to success. To help you address this, we present five common planning issues that we observed as patterns impacting many community food forests. From an ecological standpoint, it is fairly easy to appreciate the complexity and detail in the placement and layering of community food forest plants, trees, and assorted other living and nonliving things. However, trying to understand the backstory about what it takes to create and sustain a community food forest on a vacant lot in a mostly abandoned section of a major metropolitan city is not as easy.

range of community renewal programs focused on urban justice. They provide local youth with creative ways to produce and distribute independent media that centers on self-expression. Many of the core members work at a local university, Rensselaer Polytechnic Institute, or have taught elsewhere in higher education. These members coordinate and lead educational programs ranging from food production and environmental restoration to community visualization and storytelling. In addition to producing and exhibiting independent media as a means of positive self-expression, the Sanctuary supports a wide range of community renewal activities such as food literacy and access, social and environmental justice, and youth education.

When we visited the Hazelwood Food Forest in Pittsburgh, it was clear to us that there is more to community food forest projects than meets the eye. No doubt the array and arrangement of edible, medicinal, ornamental, and herbal species was impressive. We marveled at the way mushroom logs, apples, peaches, pears, rose of Sharon, hazelnut, milkweed, cucumbers, asparagus, lovage, yarrow, asters, oregano, cherry tomatoes, raspberries, mint, and many other species had been carefully positioned to resemble a young forest. We were struck by the realization that while investments in natural capital were clear in this food forest, the way the site's natural stock tied to investments in human, social, and financial capital was more of a mystery.

Since then, we learned that many food forest leaders struggle with five common planning questions:

- How can we ensure a safe and rewarding experience for people who visit our food forest?
- Who is responsible for upkeep and what is the best way to organize and motivate them?
- Will our project pass muster with municipal authorities in the face of contemporary community codes and ordinances?
- How do we keep people engaged and coming back to the site now that the plants are in the ground?
- Where do we find funding or generate support that can pay for necessary equipment and materials?

Safety

Safety is an important fundamental consideration. Community food forests are often intended to serve as a refuge, and project leaders should take care to protect a peaceful state and reduce apprehension. Simply mimicking a forest ecosystem is typically not enough in this regard. Thoughtful analysis of landscape layout and vegetation management is important, with an eye toward ensuring lines of sight to common areas, multiple places to exit, and physical separation as best as possible from potentially hazardous areas such as steep and rocky sites and threats like poisonous plants and shrubs. Mapping safety issues is another important technique to manage issues such as falling limbs or to determine where a safety railing may be needed or locations where crime control may be a concern. In the spirit of permaculture design, project leaders should identify sectors of concern and refuge related to user safety and general threats that could negatively impact the food forest experience. Determining the origin and direction of potential problems can help inform planning for functional and user-friendly sites. Examining the area in and around a food forest helps to map the directionality and nature of threats and areas to avoid. As the vegetation grows and the site takes on a variable structure, it is critical to continue thinking about and noting safety issues.

A successful community food forest design creates a space where all people feel invited to enter and harvest, relax, or explore. However, building a food forest requires plant and tree complexity, which invites exploration while looking for the harvest. It is hard for a site to achieve these goals if it cannot be freely used because security measures are prohibitive. Professionals such as landscape architects or urban planners deal with these issues on a regular basis when designing city landscapes. Urban foresters and arborists help maintain city

FIGURE 4.4. The Freedom Square Food Forest in Troy, New York, is a project embedded within a larger project of community renewal and enlivenment. The site is located within a high-crime area of the city. Some tiles in the mural at the food forest reflect the faces of neighborhood residents lost to violence. Other tiles note the values the community seeks to embody. Incorporating community values and history into the tiles is one factor that has promoted reverence for this site. The mural, now five years old, has not been damaged, and the food forest sounds a message of peace as a place where safety is expected and respected.

vegetation to minimize hazards. In chapter 9 we present how these professionals and others can be allies in tackling the challenge of creating an open and safe community food forest.

Influencing community through investment in social and cultural capital can go a long way in creating a safe environment in and around a community food forest. At Freedom Square in Troy, the concert stage backdrop includes a mosaic tile mural that was designed and constructed with community assistance. An artist led the design, and neighborhood residents were invited to create the tiles and place them in the mural. Residents were asked to bring photos of loved ones they lost due to neighborhood violence and the photos were transferred to tiles with the artist's guidance. The tiles were then placed in the wall to memorialize those lost and to pass along positive messages encouraging unity. More than one hundred people were present to take part in the activity, and the mural has since been the location for peace rallies and vigils calling attention to the neighborhood transformation that residents are leading.

Maintenance

When a community food forest is open to the public, maintenance does not necessarily diminish with time. What "maintenance" looks like, means, or affects simply changes. The living components in a food forest will eventually self-maintain as an ecosystem, but the intensity needed to manage vegetative growth and maintain benches and trails in usable condition remains significant. Additional effort is needed to sustain community buy-in and stakeholder communication. Plus, activities like removing undesirable species, pruning trees, or mulching trails and repairing benches is much more enjoyable when done in a community setting where work is shared and celebrated. People who put a little sweat into a project together have something they can identify with—or even commiserate about. That said, leaders at many sites we visited highlighted the importance of making both planting and post-installation maintenance fun. People should enjoy their experience at all

FIGURE 4.5. Maintenance problems due to fallen fruit such as these mulberries near benches or gathering spaces is an example of something that can be avoided through thoughtful choices about where certain species or infrastructure are placed. Harvesting ripe fruit before it falls and creates a mess is an important part of community food forest maintenance.

stages, which can keep them engaged. Camaraderie develops, and the group can develop a bond like that of an intramural sports team. Events should be celebratory and include activities such as eating together. Calling an event a "work party" sets the stage for a different experience when compared to "scheduled maintenance."

We found that sites with consistent participation typically scheduled work parties on a recurring basis, deciding on and publicizing dates well in advance. Websites and social media helped by listing announcements and public calendars; email listservs or newsletters are another option. Establishing a regular communication schedule with stakeholder groups in different social system zones helps them stay up-to-date on changes and see the relevance of the community food forest in their lives. Determining a contact person within the food forest community to act as the liaison makes it easier for stakeholders to know who to contact for important announcements. There might also be some people outside of the dedicated network who occasionally tend the site but are not active members within the first or second social system zones. Having a kiosk or notice board on-site where announcements can be posted is important for communication outside of the main network.

Many community food forest leaders stressed the importance of site preparation before workdays. Partially digging large planting holes for big trees in advance and laying out flags to mark planting configurations improves work efficiency and helps make effective use of volunteer time, minimizes confusion, and prevents major missteps. An example of one such misstep occurred when a site's water pipe was damaged during planting. Repairing the pipe cost more than the planting stock purchased for the workday. In other cases, new plantings were not properly guarded or marked and ended up being girdled or cut deeply during grounds maintenance.

Public Policy

There is a lack of explicit community food forest policy in the United States, at any level of government. In some cases, general municipal policies explicitly prohibit some or all aspects of community food forests. For example, it is not unusual for municipal policy to prohibit planting edible perennial tree species on public property. One of the main questions many community food forest project participants face early on is how and what local policies, if any, need to be researched, changed, or developed in order to allow the project to go forward. In *Public Produce*, Darrin Nordahl states, "Regardless [of] if the space is truly public or only semi-public, municipal government is going to have to play a leading role. Programs, policies, funding strategies, and maintenance regimens of any urban agriculture endeavor will be difficult to implement and sustain if the largest landowner in the city is indifferent."[3]

Sometimes municipalities are not indifferent but are preoccupied with the risks associated with edible tree species in public places. Early preparation to mitigate these concerns, be it through safety planning or by demonstrating success in other areas such as civic benefits, can be effective. Many agencies and elected officials consider edible tree projects as hopeless mess makers in community settings. But some are starting to reconsider and see food forests as a desirable development pattern because the plantings are confined to well-defined areas that people enter by their own choice. Overall, it is important to look into local policies that may affect or even prevent your community food forest from becoming a reality or, worse, force a change midcourse, once a project has started. Seeking out ways to work with local officials to lessen concern or think creatively about how to advance your aims in the face of preventive policies is prudent when initiating a project.

In chapter 1, we described the work the organization Fallen Fruit has done to create a community food forest in a public park in Los Angeles with funding from the City Arts Commission. Since that time, Fallen Fruit has worked with city council members to begin changing policies for edible tree species, and at least one policy has been revised to allow planting edible species in parkway swales. While this and other stories are inspiring, it is wise to recognize that the process of policy revision is slow and can bog down in bureaucracy, diverting energy and attention from other work. Project leaders must be careful about this angle and distribute labor accordingly.

The planning phase for a community food forest on private property or a vacant lot seems to have a shorter planning horizon than those located on public property, for which planning may stretch out over three years or even longer. Once a design is agreed upon and local authorities approve it, site establishment often happens in stages over a couple of growing seasons. Once plants are established in their intended locations, the following three to five years require lots of effort in the form of mulching, weeding, and pruning. During this time, the return in the form of harvest will be minimal from many perennial species, particularly the better-known fruit and nut tree species.

This is a crucial time to focus on how to maintain project momentum and enthusiasm from volunteers as well as local officials who support the community food forest. Whether large or small in scope, community food forests include a political component and it is wise for project leaders to keep a watchful eye on relevant local politics and policies. One strategy is to plant a variety of annuals and quick-fruiting perennials at the site along with the slower-maturing species to minimize the waiting time for returns on investment. This can help defuse any surge toward unfavorable policy changes and politics by quickly demonstrating the benefits of the community food forest. Planning plenty of activities at the site, whether for learning, work, or socializing, during the initial years after planting also helps steward positive attitudes.

Community Connection

Many community food forests established in the past five or six years are experiencing growing pains. The struggle is primarily related to maintaining volunteers and sustaining active community engagement. In our research, we heard many inspiring and remarkable stories of creative effort to ensure progress. One point to keep in mind is that although the site design of your food forest project is a central representation of its starting point, the extent to which you follow through on details of the design during installation will be less than perfect because of changes that inevitably occur. Those changes will include growth and decline in volunteers, thriving and dying of plants, emergence of new ideas, and abandonment of older ones.

Most projects are designed with community input, which is best, but at the same time many do not turn around and adequately communicate the nature and vision of the community food forest to their neighbors and elected officials once it moves forward. At some sites we visited, we had trouble locating an illustration or diagram of the site layout or information about the project's mission and ground rules. This means that someone newly experiencing the site or interested in being involved would be at a loss to learn about activities and programs, and even who to contact.

As we have mentioned, clear and consistent communication is critical for inspiring change through novel approaches. If folks do not know about or understand something new in their local environment, they are much less likely to use or help maintain it. Not everyone living near the site will be on board with the idea or see eye-to-eye about its meaning in the context of their neighborhood. Hosting an open forum to hear concerns and talk through them goes a long way in creating trust and diminishing negative assumptions. It is also crucial to include signs at the site with explanations of features, names, and uses of plants and what is edible or not. Provide ways for people to communicate with each other: notes at a kiosk, posted bulletins, or even a chalkboard. A community food forest that does not share its message as broadly and frequently as possible starting on day one is at a disadvantage.

In Davenport, Iowa, the Quad City Food Forest began publicizing their project from the beginning by involving the mayor in the groundbreaking celebration and working with local media to ensure coverage of the project. They also held an Earth Day fundraising concert and posted signs at the site advertising free food and asking people to respect the property. Some neighbors were concerned about increases in stinging insects and the potential for an uptick in rodents, but the site has been maintained, and by all accounts the neighbors respect it. Quad City Food Forest also used an excellent community engagement strategy by allowing families to adopt trees at the site.

Unfortunately, after three years of existence, the site was heavily vandalized in 2017. Around thirty fruit trees were damaged when someone drove a vehicle through the site, broke trees in half, ruined plant cages, and poured tar on multiple trees. The incident was publicized in the newspaper and on television. A call for support was sent out and the core group rebounded by using the event to strengthen their ties and persist, which led to additional volunteers stepping up to help repair the site. The message is that communication matters, perhaps even more so in the face of setbacks. Because a level of community association was consistent, the Quad City Food Forest team was able to rally and sustain the site.

Communication is a never-ending need, and it can be tiresome. Creating a strong and vigilant network throughout the community can help; be sure that people know who to contact to report information. Even with the best planning, unanticipated events might happen that will change the project's course and potentially alter its design, but with good communication and a strong social network, resiliency becomes a natural mechanism of a community food forest.

Maintaining Funding

Government budgets are constantly strained and funds for programs for community food forests may need to come from sources such as nonprofits, corporate financing and donations, or other philanthropic enterprises. Currently many community food forests begin by receiving a one-time grant. It is hard to say no to start-up support but when pressing forward, it is imperative to plan for follow-on support, whether from external sources, such as other grants, or self-sufficiency by seeking nonprofit status, selling products from the site, fundraising through donations, charging to attend programs, or some combination thereof. Unfortunately, it is not uncommon for many types of community projects that begin with a rush of energy and initial external support to fall apart only a year or two later due to lack of financial planning. Whether in-kind or direct dollars, long-term support for community food forests can be difficult to secure. Keeping the attention of your community can also be challenging, but it is not impossible. We recommend following the suggestions we present throughout this book for building a strong support system that includes a variety of people and keeps them engaged, because that kind of support often leads to additional funding. In our experience, luck and a bit of magic also play a role in sustaining momentum, but some structured planning can pay off.

CHAPTER 5
Planning to Create Change

Supporting civic progress and change is an important goal of most community food forests. To have a realistic shot at success, project leaders must begin planning for this change as soon as possible. Community food forests are almost always viewed as catalysts of progress and this is a powerful rallying cry, but the complexities of social change can lose out in the quest for simple persuasive messages. These basic ideas can motivate people during the early phases of development but lack the depth to inspire people to achieve long-term goals. Developing a theory of change for a community food forest can help a group of people carry out a more thorough and comprehensive planning process. It can help stakeholders agree upon standards and specific and unique designs that will make meaningful and sustainable change more likely.

Engendering comprehensive change requires a different type of thinking that includes planning processes and concepts that are specific to a community. This type of work is time-consuming and challenging, but it can have a lasting effect because of its precision and detail. Successful planning for social change depends largely on well-defined goals for positive progress in a community, and someone has to spend time thinking about how the food forest reflects the character of a community and fits within the change that those living there believe needs to happen. In this chapter, we provide a process that can help you develop a theory of change for your community food forest. First, though, we will cover some of the basics.

To begin, think of a theory of change like planning a road trip. It defines the schedule and means of transportation, lists who will be on the trip and what they will do along the way, and recognizes the hopes of all on the journey. Important to note is that although the final destination is important, it is not the primary concern while developing a theory of change. Rather, testing a destination is a first step, but most of the process focuses on what will make the trip to that point meaningful, memorable, and fun. Developing a theory of change for a community food forest requires defining short- and long-term objectives as well as conceptual and practical approaches, and then using back planning from your desired end goal to create the road trip plan. The value in following this approach is that stakeholders will be able to identify, evaluate, and defend their assumptions about change and then share their logic about the project's purpose, plan, and performance.

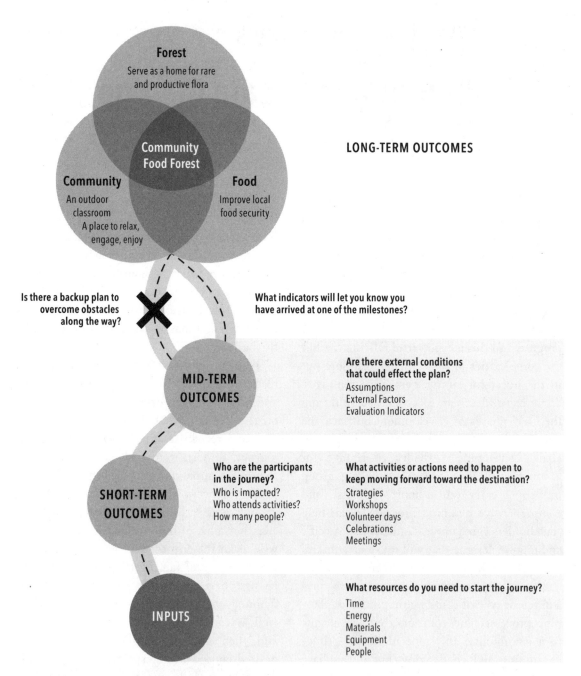

FIGURE 5.1. A theory of change begins with the vision for what type of change will occur. Rather than focus on the destination, think of why you are taking the trip to the destination. Then plan in reverse order to define the actions and short-term outcomes needed to move along the path toward the long-term vision. Two of the most important aspects of the planning are to identify your assumptions about the places you plan to arrive before the final destination and decide on what indicates you have reached them.

CREATING A THEORY OF CHANGE

How does one begin to develop a theory of change? First, it helps to think of a theory as a playbook of future food forest activity and community change. It is essentially a proposal about how the future will unfold, and no special skills or training are needed to create a plausible theory that will help define a course of action and minimize mistakes along the way. The theory of change you develop for your project will force you and others to work backward from long-term goals and identify the actions that are necessary to realize community change alongside your work to plant and manage a healthy and productive food forest. Whittling down the focus of a project to its fundamental conditions and assumptions is helpful for planning activities that are more likely to achieve desirable long-term outcomes.

What are the specific steps that you and others can use to determine and communicate a theory of change and take action to make it a reality? We offer a set of guidelines below, but best practices and experience suggest it is important to begin not with predetermined steps but with clarification and general agreement on realistic and achievable milestones. However, this is not as easy as it sounds when it comes to community food forests. Instead of creating a project under sweeping goals with little detail about true meaning, a theory of change challenges what community really means in the context of the food forest. It forces planners to identify realistic aims that are truly representative of the community where the project is located rather than what a few in power think they should be. Through a process of critical design, practical steps are then identified based on these specific goals backward to the point of initiation in the planning phase.

By working in reverse, outcomes are not taken for granted but critically challenged before being included in support of previously determined steps. First then in creating a theory of change is the need to clearly define where you would like to go. Using the road trip analogy, it is generally important to select the destination before investing time in planning. A well-crafted message about the destination can work wonders when getting a project up and running. According to the Center for Theory of Change, working in reverse is helpful for connecting dots between preconditions that are required for achieving key milestones. This somewhat counterintuitive mapping process is useful in formulating separate goals that can invigorate certain stakeholders at critical times and do so in a way that always ensures realistic and interrelated steps are identified.

Beyond backward mapping, a theory of change for a community food forest includes explicit recognition of assumptions and scheduled work whose details incorporate threats and opportunities related to these assumptions. For instance, guessing about or scrambling for equipment later in the process because availability was assumed, not coordinated, could be catastrophic early in the life of a project. Another example is the all-too-common mistake of supposing that everyone in the community will "get it" and then becoming frustrated later and giving up when many do not. It is important to explore all known assumptions because any threat of rejection from the stakeholder group due to unvoiced expectations can lead to reticence and ultimately undermine the project's legitimacy. All involved need to know what assumptions various individuals hold, whether it be expectations about availability of equipment or general acceptance of the project. These assumptions need to be critically discussed to avoid technical misfires and community confusion.

From there, the particulars of management can be outlined as specific tasks such as an opening event, fence raising, educational training, maintenance days, and other activities that stitch together steps defined during backward planning. Though difficult, this detailed planning can be a little easier for stakeholders and community members beyond the nuclear group to digest and understand, because it concerns tangible tasks rather than concepts and themes. An example of this is the decision of the organization that founded the Dr. George Washington Carver Edible Park in Asheville to change its name from Edible Entropy to City Seeds, because the former was so nebulous it created confusion rather than community. In general, as projects mature, telling people in a practical way what you are going to do usually produces better results than speaking conceptually.

Many involved in theory of change work recommend creation of a document that spells out the project's final aim, anticipated milestones and their connections, and assumptions and how they relate to particular interventions and work along the way. These interventions are also shared in as much detail as possible. How would this document be used? One example is a stakeholder looking to market the community food forest to audiences that want to know what they can do to help. Having an itemized and well-defined list of activities needed to move the project forward can pay important dividends in the long run. Or, if you are working with a group that seeks to understand how your project fits into a larger narrative around change, perhaps for purposes of writing a proposal or lobbying for political change, then the document's content related to milestones and mission could be very useful.

Creating a theory of change is fortunately not left to each project to start from scratch. Groups such as the Center for Theory of Change and Research to Action provide free online software and a set of published questions to ask for each outcome. These resources and others like them can be used to develop indicators and leverage points that help community food forest stakeholders build on the general description above and create their own theory. (Check www.communityfoodforests.com for more information.) Regardless of the specific content, all theory of change documents will need to be a living work that can adapt to new situations and integrate emerging ideas as the community food forest evolves.

The Role of Whole Measures

Some of the questions asked when developing measurable outcomes in a theory of change can be intimidating because they require decoding intentions and weaving in accountability, which may come across as questioning the integrity of good intentions and unnecessarily delaying action. For instance, a question such as "Is the time frame realistic?" could be misconstrued as an attack on someone's ability to get something done. Or asking questions such as "How does your specific context shape your outcomes?" might create debate about assumptions regarding community culture. However, it is wise to take time to think logically rather than emotionally and chart strategies before jumping into a project. Understandably, some organizers are concerned about momentum, yet framing logical thinking as a reflection of the values that are at the heart of a community food forest can also create stamina that is helpful in working through the process.

The Center for Whole Communities outlines a Whole Measures framework for shaping strategies that include diverse perspectives.[1] Whole Measures is a good starting point for reflecting on what values most readily resonate among community food forest stakeholders and are likely to remain relevant over the life of a project. If decisions are difficult, and many will be, it is helpful to think in terms of Whole Measures as a way to narrow down and choose values that most strongly relate to a project. Whole Measures shifts the idea of measurement from tasks to domains of impact that nurture positive relationships within and between communities and the environment.

The Center for Whole Communities outlines key constituents of the Whole Measures process. Included are concepts such as narrowing and aligning priorities, finding ways to inform dialogue and planning along the way, strengthening collaboration through priority alignment and informed processes, and developing useful and meaningful evaluation and outcome reporting programs.

Difficult questions arise at the outset of any community food forest project, such as what will change in the community and when, who will the food forest reach, and how to begin the conversation about theory and application. For example, if community building or collaboration is consistently discussed, then questions would follow about the details of relevant community audiences that can help determine what to build. If this means building relationships with local neighborhoods, then metrics for success could be fairly straightforward and include, for example, the number of families that visit the site within two years of establishment. Or metrics could be more holistic, such as extent of community recognition and developing a sense of place. It could also mean that a balance between community residents and technical experts is necessary to make decisions during the initial planning and design period. Other relevant examples include considerations such as whether a translator should be present at meetings. Does an anonymous polling system need to be planned for meetings or should surveys be distributed to stakeholders? What formal structure is in place to allow for and manage disagreement? How will intragroup and user conflict be handled?

Deliberation about community projects leads to better design but is challenging and can be time-consuming. Thinking through what and how a project hopes to benefit a community before proceeding increases the likelihood of an equitable and impactful community food forest. A theory of change for a particular site is its backbone and perpetual undercurrent. The method is generally straightforward and can be accomplished with publicly available tools. Whole Measures refines the process by accounting for community diversity and working to engage and inspire.

OCEAN VIEW GROWING GROUNDS

The theory of change and whole measurements helped the Ocean View Growing Grounds food forest in San Diego, California, develop their community garden and food forest project within a bioregional sustainability program focused on social change and improving relationships between communities and universities. The idea for the food forest began with Keith Pezzoli. As director of the University of California at San

Diego's urban studies and planning program, much of his work focused on urban sustainability, green infrastructure, and project replication. He was especially interested in neglected areas where pollution is a concern affecting urban agriculture, food access, and public health.

In 2009, Keith used asset mapping to survey southeastern San Diego. The asset survey focused on vacant lots that could be transformed into places for growing food and perennial vegetation that also could help protect watershed health. After finding more than 800 vacant lots in his survey, Keith began thinking about how such space could be used to improve local food security and sovereignty. If neighborhoods could see these places as assets, then some might use them for activities such as food production and community celebrations. Serendipitously, Bob Georgiou, an alumnus of the UCSD urban studies and planning program, owned one of the vacant lots identified in Keith's analysis. It was used as a local dumping ground by some and contained several abandoned vehicles and other waste that Bob had to remove periodically.

Checking Assumptions with Community Input

Being involved with the urban studies and planning program and aware of Keith's study, Bob offered his lot for development with the challenge of transforming it from a community liability to an asset. The property was a 20,000-square-foot site located in the Mountain View neighborhood of southeast San Diego. It was an ideal place to test different forms of planting that could help clean up the lot and provide produce while helping to protect the watershed. In line with his asset-mapping project, Keith envisioned an experiment that would study and demonstrate various growing methods that residents could replicate in their backyards. There, residents could gather and provide information on how to deal with the challenges of urban production while also increasing green infrastructure in their neighborhood. Community buy-in was considered critical for this plan to work, and any development of the site would have to reflect the desires of neighborhood residents.

Bob leased the site to UCSD for $1 a year. Once the handover was complete, Keith and his wife, Janice, collaborated with a local community organization called One Hundred Strong to canvass neighborhood residents about their ideas regarding next steps for the project and how they define the boundary of their neighborhood. In other words, where the neighborhood starts and stops in the minds of its community members. They also held meetings in a celebratory style with music and food to attract more people to the site for discussion on next steps. In this way, neighborhood residents were deliberately included from the beginning, which proved very useful in developing a theory of change based on community input.

The site was named the Ocean View Growing Grounds, and after considering various options for creating a permanent community garden, project leaders decided to incorporate a food forest because it would be perennial and diverse. Also, the food forest seemed like a good way to satisfy multiple concerns voiced by residents and offered the possibility of decreasing maintenance needs over time, while also producing local fruits and nuts and providing ecosystem services such as storm water retention. It would contribute to the goals of long-term land rehabilitation as well and help neighbors learn about and replicate similar food forest projects in their backyards.

Defining the Vision for Change

Keith was sensitive to the fact that universities and their representatives are sometimes seen as community users rather than partners. He wanted to work through this project in a way that dispelled this perception because he believes access to science and technology are basic rights. To shift community perspectives and reduce discomfort regarding the nature of universities, Keith created a community-based organization called the Global Action Research Center that would be the point of contact and conduit for decision making about the site. The nonprofit organization would help put university research into action while at the same time act as a liaison between the neighborhood and university about site use and research results. In doing this, successful aspects of the project could be distilled and then transferred to other neighborhoods. However, ideals and ambition led to an organization that tried to be everything at once. Managing such a dynamic and comprehensive program became too much for one person, and Keith eventually joined with Paul Watson and Bill Oswald, both former executive directors of community building and social justice organizations that are inspired to lead social change via popular education. They were versed in using the theory of change and brought it into the Ocean View Growing Grounds project.

Popular education is a process of critical analysis that identifies the primary issues affecting a community. It helps inform people so that they can take responsibility for personal contributions to achieve desired outcomes. The goal of popular education is informed action that changes the accepted "norms" about a situation by calling attention to underlying causes and designing interventions to create preferred transformations.[2] Keith and Paul focused on supporting people in their efforts to address critical issues by learning why the issue exists, identifying the social structures and political decisions that impact them, and questioning imbalances of power that affect communities. Paul is now director of the Global Action Research Center and regularly uses theory of change to design interventions that strengthen neighbor connections and increase their command of the research and technology at the University of San Diego.

Rather than limiting the effort to create a community garden with residents, both Keith and Paul used the lot transformation to ensure all community residents had a say in defining their food network and issues related to it. The hope was that people would question why their community is viewed as a food desert, why and how such areas exist, and what education and action is needed to empower self-determination in this regard. Their theory of change focused on ways to create healthy and secure connections between people and their community land as a necessary prerequisite to civic engagement and change. The Ocean View Growing Grounds would create social bonds and reconnect neighbors to the land while providing a place for education, self-governance, and a local food network.

Using a theory of change to define the vision for how the site will support social progress helped chart a particular path forward for the community food forest. The process helped determine the conditions needed for preferred change, along with metrics for evaluating progress toward goals. The project was not simply a side project of the Global Action Research Center. Rather, the community food forest became a central part of the overall program for change, not only in the local food system but also in the civic relationship between a university and a community. It grew into a place where people and institutions interacted in new and exciting ways through a model of citizen rights related to the design and development of research and renewal. The vacant lot became a place where ideas and action fed a living laboratory that tests assumptions, supports innovative practices, conducts research, and delivers inclusive educational experiences.

Short- and Long-Term Inputs

The Ocean View Growing Grounds provides multiple functions that reflect its many stakeholders. Paul had extensive experience working with low-income and distressed communities and knew that most people understand their problems very well and have innovative ideas about solving them, but rarely have adequate resources to do so. Thus, he was very keen on leveraging partnerships to help connect people and resources. For example, Bob Georgiou is now on the board of directors of the Center. As a way to give back to the project that helped transform his property to a useful site, Bob occasionally provides financial support when materials are needed.

The evolution of capacity building at Ocean View Growing Grounds occurred over the

FIGURE 5.2. The food forest at Ocean View Viewing Grounds is used as an educational site for teaching community members how to replicate practices on their own property. Raised beds demonstrate one method for building soil health and mitigate uptake of any toxins present at the site. The beds also offered an opportunity for those involved to decide democratically how to share the growing space.

course of several phases, each with their own lessons learned and interesting adaptation. Early work focused on building community relationships and conducting a site assessment. Funding from an EPA Superfund grant to the university supported extensive soil testing for potential toxins and promoted ways that local residents could replicate environmental management at their homes. Community members and UCSD students helped with the soil testing, and raised beds were identified as one preferred solution to the challenges of brownfield planting. Residents used an agreed-upon governance structure to manage the garden and decide who would use the raised bed plots. A researcher also initiated a study of heavy metal accumulation in trees on the site.[3] The study aims to quantify and monitor heavy metal accumulation from the site and assess toxins in the food forests. Here science and technology directly benefit the health and safety of the neighborhood and help all involved understand how to create functional plantings in a place of environmental and social importance.

Initial development of the community food forest at the Ocean View Growing Ground started with resident-led fundraising and donation of materials such as old fencing that would be useful for developing and maintaining the garden. A grant awarded by the University of California's Global Food Initiatives program funded the living laboratory portion of the project, which included bringing in design professionals to collaborate with residents, planting trees in two sections of the food forest, and installing an irrigation system. The Center along with partners organized design charrettes with residents to create a master plan for site layout and infrastructure so that input from all stakeholders shaped the development of the project.

Short-Term Outcomes and Long-Term Impacts

The site intentionally included two main components that were created to meet the needs of all stakeholders: the Learning/Action Center and the Neighborhood Food Network hub. The site was designed with community garden plots, two strips of multistory edible perennials bordering the front and back of the site, and a green gathering space where workshops and hands-on demonstrations are possible on topics such as gardening, nutrition, and produce seasonality. The Learning/Action Center of the garden and food forest were recently completed. Next steps include expanding the Neighborhood Food Network, all of which has been created by 161 residents. The long-term goal is to increase this number and reach as many of the 500 or more neighborhood residents as possible.

The site has far-reaching objectives focused on connecting and empowering neighbors through food and ecosystem services to improve health, recognition, and functionality of a once underutilized space. Short-term change is already observable through site transformation, educational workshops, and connections between university students and a community service research site that helped develop stronger ties to the local community. Next steps toward longer-term change include asset mapping with the residents currently in the Neighborhood Food Network and then expanding to other areas to determine who has usable space on their property and is willing to include it in the network. Additionally, maintenance skills and food production experience among network members are being mapped to create a community network. The data will inform food-growing techniques taught at the Learning/Action Center so they can be replicated by residents.

When mapping is complete, members of the network will help determine what space can be used, what will be grown there, and who will be involved in maintenance. Plans are to canvass the neighborhood twice a year to increase membership. Additional aims include regulating maintenance and continuing to publish the bimonthly newsletter, which helps the network communicate with each other and stay informed and invites others to join.

CHAPTER 6

Rooting in History

There is no question historical events profoundly shape current conditions and how a particular culture and place have developed socially and ecologically. Community food forests reflect a history of planting, tending, and managing forest ecosystems for human use. At the same time, they play a role in changing models of civic progress and shape how we understand and envision a resilient future in built environments. They represent a crossroads where traditional practices meet modern approaches to bring forth innovative ideas about blending social and ecological spaces. It is for this reason that investigating site history before beginning a food forest project is useful: such context helps anchor a food forest to the stories, history, and pride of its community.

Thinking objectively about the history of food forests and how a project finds firm footing in your community can help answer questions about the issues or assumptions that are likely to affect current efforts to design, implement, and promote a project. Reflecting on contemporary food forest trends enhances your ability to work effectively with partners and stakeholders toward implementation. Finding a place for relevant civic issues will lead to a culturally informed and historically grounded project that not only helps connect people to land and makes the neighborhood, town, and city greener but also leads to meaningful intersections between local action and broader issues.

Here are some key questions that can help guide efforts to align your community food forest project with the historical and cultural contexts of your community. Of course, there are plenty more to ask, but these will help you get started in tailoring your efforts to balance local history and culture adequately with the exuberance and excitement of the community food forest movement.

- Where, if anywhere, are realistic community food forest sites located?
- Which plant and tree species do potential food forest sites favor or preclude?
- What is the social and environmental history at these locations?
- Who are the community members who will use the food forest?
- Who are the community leaders who can help support your project?
- How and to what extent are people in your community truly disconnected from land and food production?
- What does it mean in your community to be disconnected from land and food production?

- Why and when did this disconnect happen, if at all?
- In what ways will the public value your food forest?
- How will your food forest be sustained and community well-being improved?
- What are the political steps and institutional measures you should take to ensure success?

REFLECTION AND INSIGHTS

Branding and promoting a community food forest project by relying on generalizations about greening and growing are not likely to lead to long-term accomplishments and change. There are many more relevant local issues and interests that stand to pay significant dividends for projects. For example, investigating historical meaning and incorporating that meaning into a project is more likely to lead to broader community awareness and interest. Practical initiatives rest squarely on this sort of homework. Digging into complex civic issues, historical precedents, and site characteristics may seem relatively simple but can be time-consuming and require significant dedication, especially if site leaders do not live in the neighborhood.

So how does one assess and map their community's biophysical and social context? How can these insights be used to design and manage a food forest that engages and wins over stakeholders? And how is this accomplished while grounding work in deeper meaning shaped by history? One place to begin is to learn about the history of community food forests overall, and in this chapter, we cover some of the historical aspects that have influenced our contemporary notion of community food forests. Our focus is on the Western perspective since the community food forests we visited were located in the United States.

Although one perspective is generally offered, project leaders are strongly encouraged to consider history as a story with many relevant versions depending on culture, race, ethnicity, country of origin, and other innumerable factors. Key characters, important milestones, and ingrained patterns of belief will undoubtedly differ. Every story matters and is relevant to the development of a unified community project. Also important are how food forests influence social perspectives regarding community environmental stewardship. In the end, history influences how we think about community food forest objectives and the ways they are linked to larger social movements emerging in the last few decades.

PERSPECTIVE

Is the use of public land for perennial food production in a community park something new in the United States? Not really. Public places have long provided opportunities for community-based perennial food systems in the United States. They provide green spaces and places for people to pursue diverse aims, be they civic, financial, environmental, personal, or something else. But in what regard do people hold green space in their consciousness?

When asked, most people would probably say they consider green spaces useful, but it is safe to say that most do not think of public perennial food access in terms of larger trends of American land use and environmental history. Yet in many cases, these spaces and their resources have a historical precedent and philosophical purpose established decades earlier. However, detailed knowledge of the history of food forest production is probably quite rare even for those who are actively involved in establishing community food forests.

Food forests are a modern-day version of the intuitive land management techniques practiced long before colonists arrived in North America. During colonization, the use of many communal indigenous land management practices such as food forests dramatically declined due to cultural changes and the decimation of the Native American population from disease. Following the colonial era, rapid industrialization and a growing population drove US farms toward monocrop production because it could produce enormous yields. However, monocropping also requires vast capital and intensive mechanization to realize such remarkable output. The result is a food system where the vast majority of people, places, and production are largely separated and compartmentalized instead of integrated. Sustained production of these intensive systems requires immense inputs primarily derived from fossil energy and dependent upon chemical pesticides, and there are grave concerns about the long-term environmental impact and sustainability of such production.

Thus, there is growing recognition and acceptance that collapsing scale for some of our food production can play an important role in feeding the future. The interest in community food forests runs parallel to movements in community-based and local agriculture and is derived in response to a growing sense of estrangement from food production. Alternative, localized connections between consumers and producers of food are on the rise. As described in chapter 1, the amount of direct-to-consumer farmers markets and registered community-supported agriculture programs have dramatically increased in the past two or three decades, and notable segments of participating producers are using polyculture design and cultivation rooted in agroforestry, permaculture, and agroecology.

THE BOSTON FOOD FOREST

One excellent example of the coevolving nature of deep historical meaning and contemporary expression is the food forest initiative in Boston. The area is steeped in history. A sense of pride emanates from its strong traditions and stories. The first university of the colonies and now one of the most prestigious in the world, Harvard University was founded there in 1636. It was only about a decade earlier that apples came under cultivation in New England, after which Boston became an area where hundreds of varieties were grown for many uses. According to Rowan Jacobsen of the *Boston Globe*, "Apples were some of the leading protagonists in the story of American ingenuity, diversity, and prosperity."[1] Varieties were grown throughout America, and many were regionally acclimated and reflective of cultural preferences in settled areas. Yet today most people have little idea that cultivating apple varieties was once a common pastime among everyday people.

The Boston Food Forest Coalition (BFFC) is forming a network of neighborhood-based food forests and reviving Boston's legacy orchards. The core group leading the community food forest project met through a Transition Town network that formed in Jamaica Plain, which was originally settled for farmland and is now a four-square-mile Boston suburb. It is an ethnically diverse and highly desirable neighborhood with the most per capita green space in the city. It is also located along a system of meadows, waterways, parks, and botanical gardens known as the Emerald Necklace. Frederick Law Olmsted, a famous landscape architect of the nineteenth century perhaps best known for shaping our modern city parks, designed it as a ring of green space to cure unhealthy urban conditions.

In 2009, three Jamaica Plain residents who happened to connect through their work with the Institute for Policy Studies shared their concern about the economic and environmental state of their neighborhood. They scoped out projects to build community resilience and address issues in their neighborhood. One action was transforming vacant and trash-filled lots into orchards. They also created forums to foster dialogue on the state of the community. Orion Kriegman, one of the residents, would later become the executive director of the BFFC.

Orion and others cofounded the Jamaica Plain New Economy Transition (JP NET), which became a way for residents to connect and offer support to one another. It also provided an opportunity to connect to larger movements through the regional New England Resilience and Transition Network and the national Transition Town Movement. The Transition Town Movement (sometimes referred to as the Transition Initiative or Model) is "comprised of vibrant, grassroots community initiatives that seek to build community resilience in the face of such challenges as peak oil, climate change and the economic crisis."[2] JP NET became the ninety-fifth official Transition Town in the United States. The Transition Town movement inspired the initiation of at least three other community food forests. One of the JP NET programs was the Community Leaders Fellowship, which provides its fellows with coaching, training, strategic planning, fundraising, and other support toward leading innovative change. Daniel Schenk and Allison Meierding, both core members of the BFFC, were fellows.

Through the BFFC, Dan, Allison, and a few others secured permission to install the flagship food forest site on the grounds of Mass Audubon's Boston Nature Center. The site offers educational opportunities to the city's flow of visitors through its youth programs. It is located near the Clark-Cooper Community Garden, which has 250 plots on eight acres. The garden is the oldest and largest in Boston with a long history of traditional gardening methods. Longtime gardeners at first expressed skepticism about permaculture practices and the intention to incorporate them into the food forest, but much of their concern was allayed through exposure to permaculture demonstrations in the food forest. Techniques such as a hügelkultur and herb spirals were showcased and helped minimize the perceived risk among gardeners about implementing such practices in their own plot. One of the food forest leaders gained

Looking Backward and Forward for Social Change

Contemporary issues and social movements influence community member action. They help bring together people who share interests in social change. The Boston food forest site added local context by demonstrating food forest practices and uses to community gardeners and working slowly to increase acceptance. Historical context can also be used to enhance a project's meaning and relevance. This information helps provide a compelling story that site leaders can use to gain community interest in the project.

credibility by tending a plot in the community garden and validating her gardening skills by producing an abundant crop. Interest then began to grow in permaculture methods, and she used her new credibility to bring some of the gardeners to the adjacent food forest to observe firsthand the impressive yield in the hügelkultur area, even during drought conditions.

The Massachusetts Audubon Society (Mass Audubon) acquired the property on which the food forest is located in 1997. It was a state hospital for decades, and much work was needed to remove buildings, restore land, and build the Boston Nature Center (BNC). BNC staff were curious about the site history, particularly Steven Pavlos Holmes, who was a scholar in residence. Steven sought to find answers to questions about how the land has affected the people who have lived, worked, and played on site. The result of Steven's research was a book, *A Healing Landscape*, which tells the story of more than two centuries of transformation and use at the nature center site.[3] Daniel Schenk has acknowledged that the site's history as a healing landscape is the core of his motivation to create a food forest that is a restorative place for both the land and people.

The Massachusett tribe of American Indians lived in the area before their existence was upended by the arrival of Europeans. The Massachusett gathered wild fruits such as elderberry, blueberry, raspberry, strawberry, and blackberry. Nut trees such as pignut hickory and oak dotted the landscape along with useful herbs like boneset. These native species were historically important, providing critical nourishment to the people residing in the area.

After the European arrival in the 1600s, the marshes, forests, and meadows of the area were converted to farmland and divided into multiple parcels throughout the 1700s. Parcels that made up the land of the BNC stayed within the hands of a few families into the 1800s. There was a strong familial connection to the land as home, provider, and place of natural refuge and quietude. In the 1800s, the natural areas were so beautiful that they inspired some of the writings of Ralph Waldo Emerson, who was a young school teacher in the area at the time.

In the late 1800s, after the area was incorporated into the city of Boston, the landowners sold their property to the city, which then used the grounds for a psychiatric hospital. Over time, adjacent farms were bought and added to the property. At that time, new cognitive health care theories referred to as "moral treatment" were emerging. These treatments called for removal from city life to rural places where patients could benefit from fresh air, exercise, and hard work. Patients worked alongside long-term dedicated hospital and grounds staff to keep the farmland in working condition. Records of crops grown on the hospital's farm demonstrate a blended approach of combining annuals with perennials. Some of those perennials, such as apples, currants, pears, plums, raspberries, cherries, strawberries, asparagus, dandelions, peppergrass, and rhubarb, can still be found in the food forest site today.

The BFFC was able to draw on this rich history to relate the objectives of the food forest to community members using historical and modern stories. Making such connections is critical. For one, they reify the importance of social capital, which can create access to networks that increase possibilities for acquiring land, funding, or other resources. Additionally, they can enhance networks among those who are in a position to influence or promote community food forests to people who can make things happen. In response to how the food forest concept was explained to BNC administration,

FIGURE 6.1. A hügelkultur and herb spiral are located in an accessible area of the Boston Food Forest so visitors can easily see permaculture techniques. They also are visible from the community garden located across the road. The gardeners were skeptical about these permaculture practices at first, likely because many tended their plots in more traditional ways. The demonstration and outreach by food forest members in the community garden led to dialogue about the benefits of contemporary perennial polyculture production.

BFFC members prepared a proposal explaining how it could rehabilitate land, connect people to nature, demonstrate sustainable practices, and tie into the history of the land as a source of perennial food and as a restorative place.

A DEEPER GROUNDING IN HISTORICAL CONTEXT

The community food forest in Boston is one small example of how history can create deeper meaning for a site and strengthen its importance in a way that calls attention to restorative community spaces. In the remaining sections of this chapter, we delve more deeply into the historical context that we believe helps illuminate why the timing is right in the early twenty-first century for society to embrace community food forests. We cover important milestones that have prepared the collective conscience for accepting and supporting the idea of growing perennial edibles on public land. You may find this context helpful in framing the history at your own site or, in a more general way, winning interest among peers and the public in the food forest concept.

Convergence—1600 to early 1700s

When Europeans and North Americans converged in large numbers approximately 400 years ago, communities native to the American continent were deeply advanced in terms of commerce and land use production, much of which included intensive swidden agriculture and hunting and foraging across broad landscapes. Though not without debate, what can be generally gleaned from historical documents is that the North American diet was largely based on edible perennial species, game meats, some animal husbandry, and cultivation of some annual plants. It is also known that health care was based principally on diet and use of particular medicinal plants, guided by community members with divine spiritual intuition and primary care skill sets passed down through generations.

Of course, much changed during European colonization, including violent conflict and tragic indoctrination, assimilation, and social control. In *Rural Communities,* the authors explain community capitals and their relevance to rural communities by pointing out a distinct difference between European and North American paradigms: "Whereas most Native American tribes used the land to develop a subsistence economy, with a strong focus on converting natural capital to social and cultural capital, most Europeans came to the Americas to transform natural capital into financial capital."[4] In this regard, Europeans crafted a new nation by altering the physical and social landscape to support commodity crops and extraction of timber, animal pelts, minerals, metals, and other marketable goods.

Part of European hegemony included increasing fixed footholds on land by fencing off parcels of property in a way that was largely incongruent with precolonial perspectives and practice. Much has been written and debated regarding the extent and impact of this shift in cultural lifestyle and land use in the Americas, but perhaps most relevant to consider in terms of contemporary community-based perennial food production is the contemporary shift toward using natural capital for social and cultural benefits rather than only as a means for financial gain. Also relevant to food forests is that, over time, a Euro-American melding of food production and consumption emerged wherein food production combined with polyculture plant production and wild meat harvesting to form a unique North American food sourcing system. It began that much of the postcolonial diet was based on breads, porridges, and pies made from European rye, oats, and wheat, in addition to substantial integration of native corn. Animal domestication was also increasingly popular, with pork favored in particular along with beef and poultry. Vegetables and fruits were consumed as supplements and grown mostly in home gardens.

Trees and shrubs were integrated for perennial production of fruits, nuts, and berries. Food obtained from hunting and foraging supplemented farm production. Some vegetables that we consider common today, such as tomato, eggplant, and potato were not widely accepted in early colonial diets. In fact, at that time it was not common to eat leafy greens alone except in a few European cities. Root vegetables and squashes were common, but beans and corn were "New World" additions. Vegetables were generally used to make sauces that accompanied meat and bread dishes. Herbs were grown in gardens primarily for medicinal uses. Cinnamon, nutmeg, clove, and mace were often used as curatives in addition to flavoring desserts. Apples were the most common and favored fruit, and more than 1,000 varieties were imported from Europe and grown on countless homesteads. Many apples were pressed to make cider; later, eating whole apples became more common. Although a wide range of native fruits and nuts were gathered from the wild, they were increasingly planted in orchards and groves, too.

By the mid- to late 1700s, farming in North America was increasingly based on annual crop cultivation mixed with perennial food- and fiber-producing woody plants and trees. Seasons of intense annual cultivation were followed by fallow periods, using cover crops under perennial species to renew the soil. These forms of food production helped fuel the American Revolution, which was forging an identity for newcomers to the continent just as they were transforming the landscape from forest and meadow to a hybridized agrarian setting. Out of this came a uniquely American polyculture, born out of a convergence of cultures and integrated over time to form practices that inform contemporary community-based perennial production in fields such as agroforestry, permaculture, and agroecology.

Transformation—late 1700s to 1800s

The United States was in its infancy and the Industrial Revolution was taking shape around the time America's food production identity began to form. As mechanization advanced and commerce grew, immigrants from Europe and elsewhere began arriving in great numbers, many intent on finding employment, claiming land, and building opportunities for social mobility. The country was in fast-forward. Built, political, and financial capital become dominant norms for social progress. Technological advances were rapid, economic growth was high, the population was increasing, and America was transforming from a mostly small-scale agrarian society to a mechanized, urban, industrial one, based in large part on textiles and iron produced from natural capital.

The impact of rapid population growth and increasing mechanization on food production in the United States was immense. Another period of swift evolution in its food system began, defined by pivots from small-scale agriculture, hunting, and foraging in the shallow rocky soils of the East to large-scale production based on mechanization, mass transportation, and economies of scale in the deep, fertile soils of the Midwest. The population grew and industry flourished. Cities blossomed that housed labor and corporations. Methods for acquiring capital again shifted. Previously, human capital such as skills and professional experience were acquired through apprenticeship in a craft or passed on through family traditions. Now they were specialized and privatized through bureaucratic and often costly education.

Education was critical for gaining ground in urban life, where the goal was to obtain the cultural capital of high society. Thus, urban centers were literally fed by ever-growing yields of produce from vast and increasingly efficient farms mostly located in the Midwest and South. Distance to market and volumetric processing became less and less important as transportation opened up on waterways, railways, and improved roads. Innovative mass sorting and storage systems were marks of progress and ingenuity efficiently managed at factories using assembly line aggregation and distribution.

It was a new epoch of food production and preference wherein people were increasingly separated from land use in terms of production yet fed by ever-growing volumes of more diverse food products provided year-round. Disconnection from the land increased, and appreciation of seasonal rhythms of availability for specific foods decreased. The merging and mixing of food production and cultural preference that defined the early character of colonial America was largely ushered out. People became increasingly separated from food production and sourcing, which inevitably precipitated a shift in core values, assumptions, and beliefs that permeated the next 200 years.

American society changed quickly and quite dramatically in the mid-1800s. Standards of living improved and modern-day science and technology provided new means for understanding and managing the world. But even in the earliest stages of industrial evolution, there were signs of social discontent. Though much of the modernizing world embraced existence under the moral guidelines of classical economics, some in the cities sensed a profound separation from the physical means of their existence.

Nature as Recreation

Perhaps one of the best known focal points in this growing alienation is the reaction to Henry David Thoreau's *Walden*, published in 1854. Renowned for its reflections on connections with nature, the book was based on Thoreau's experiences living in a small cabin in the woods. Thoreau extolled a natural existence and connection to the environment. The book captured the imagination of an increasingly urbanized society, cementing romantic notions of nature in the psyche of an increasingly postmaterial society. Another case is New York City in the mid-1800s where, by 1858, the upper and middle classes were demanding open space in urban centers for recreating and socializing. They were interested in making the city's green spaces comparable to those in London and Paris. The result was New York City's Central Park, which Frederick Law Olmsted designed.

Parks such as Central Park became national icons and marshaled in a period in green design history focused on "pleasure grounds." Pleasure grounds were meant to be public places for recreating, relaxing, and communing within a pastoral landscape, offering a highly fashioned escape from the hampered and health-threatening spaces of dense metropolitan quarters. Interestingly, they were inspired in large part by nostalgia for agrarian countryside lifestyles. They offered a healthy alternative to taverns and other cramped spaces by providing open and natural places to socialize, move about, and experience the nobility of community associated with rural life. Yet the simulation was reserved for only some.

Central Park, for instance, was distinctly upper class and created at the expense of Irish farmers, German gardeners, and African-Americans who lived on and worked the land at the time it was partitioned. Time for relaxation and leisure in the park, reminiscing on the aesthetic and feel-good qualities of country life, was a luxury that only the upper class could afford. For the rest, some still lived the country life, experiencing it for all its hardship, while others could rarely if ever afford to step away from the reality of their daily circumstances. The story of New York City's Central Park is one example of the cognitive dissonances and incongruities permeating mid-nineteenth-century urban life. Its rapid-fire industrial spirit and avaricious leanings were juxtaposed with a view of rural life as nobly connected to core existence through the simple values of farming and family.

By 1860, one-sixth of the American population lived in cities. Outside of structured natural escapes such as Central Park, pigs, dogs, and other animals ran around and dirtied the city streets and open common spaces, leading to laws banning domestic animals from cities. Urban agriculture was thought to be unsavory, triggering its banishment in the name of public health. The poor, of course, suffered the most, losing perhaps their sole access to meat, milk, and eggs.

Not only was the identity of the nation changing from rural agrarian to industrialized urban, but the nation's agricultural production system was as well. The slave system in the South ended after the Civil War, and the nation continued uprooting and confining native populations to some of the least fertile tracts of land on the continent. At the same time, massive gains in industrialization were underway, and Europe was undergoing a social and environmental upheaval of its own, much of which defines our current worldview with respect to the environment, be it wild or built.

Nature as Resource

The word *ecology* was coined in 1866 by Ernst Haeckel, a German zoologist, to describe the relationship between organisms and the organic and inorganic components of a given environment at a particular scale. In the 1870s, the notions of an ecosystem—a bounded ecological landscape—supported a variety of values taking root in the United States. Discovery and consciousness of ecological principles grew in the American mentality as westward expansion introduced a generation to new and profoundly beautiful and varied landscapes. One outcome of this emerging perspective was the creation of Yellowstone National Park in 1872. It was one of the first major investments of financial capital to conserve natural capital and begin investing in a culture of nature.

Interestingly, 1872 was the same year Arbor Day was celebrated for the first time. Held in Nebraska, the event promoted use of tree windbreaks around homesteads and fields to protect soil and crops, a form of agroforestry still practiced today. Reports indicated that one million trees were planted that first Arbor Day, which went on to become nationally recognized and is now the date used to celebrate Earth Day. These early advances indicate a growing sense of environmental complexity and consideration for nature beyond anthropogenic value. Before this, wild nature was considered either divine or largely unexplainable and unsafe, but now it was viewed as a system made up of parts that, because of industrial and scientific advances, could be measured, studied, explained, and managed. This shift made nature approachable, interesting, and capable of being categorized and understood. The path was not without challenges, though, and this era represents only the beginning of such socio-ecological awareness.

Agriculture over Nature

Industrial agriculture in the West, from the Mississippi River all the way to California, arrived in earnest by the end of the nineteenth century, and small farms entered a long period of slow, steady decline relative to overall population growth. At the same time, urban populations continued to increase. Packaged and prepared foods became a more significant part of the American diet. Broad access to electric power improved city life, which became increasingly attractive to rural residents who had limited access to innovation.

Electricity also opened the door for development of kitchen appliances such as refrigerators and woodless stoves, which began appearing in houses and buildings around the turn of the century. Food brands still popular today, such as Kellogg's and Post, got their start. Although refrigerated railroad cars were first developed in the mid-1800s, they had limited availability to ship perishable foods long distances. That changed in 1878 when advances in refrigeration made the slaughtering and processing of meat, along with production of other perishable products, possible many miles from where the foods would be sold.

More brands still available in grocery stores today came into being during the latter part of the century. Examples include Tabasco sauce, Fleischmann's yeast, Welch's grape juice, and Philadelphia cream cheese. The economy continued to grow, and a middle class with discretionary income fostered a new generation of consumers. New recipes emerged in response to the flood of available foods and income, leading to rapid growth in the adoption of kitchen technologies. As access became easier and time spent preparing food began to shorten, the American concept of food increasingly took on an expectation of immediacy and variety, which resulted in a great deal of experimentation in kitchens and the rise of the restaurant economy.

Yet in contrast to the growing ease of food access and variety, city life was becoming more congested and unhealthy. During the second half of the nineteenth century, the proportion of the American population living in cities rose from one-sixth to one-third, causing significant welfare challenges and placing great demand on infrastructure. Modern plumbing, along with advancements in public transportation, helped improve the quality of life for many, and high-rise buildings were constructed to house growing urban populations. Subway systems were developed to avoid the tangles of aboveground activity, with Boston completing the nation's first line in 1897. Problems related to health and crowding, however, seemed constant and unending.

Though beset with growing pains, ever-expanding urban centers drove the nation at the turn of the twentieth century. They provided jobs, offered entertainment, fueled innovation, built wealth, and defined social identity. The exciting cultural progress drew people from all corners, despite largely bleak living conditions. Perhaps because of efforts to balance green and gray infrastructure in urban design and ecological consciousness, social movements emerged that emphasized greater equity and practicality with respect to cityscapes and locally produced food. Pleasure grounds were considered useful but passé, and improving access to nature and locally produced food for all residents became the next wave of design.

There was no denying the benefits to human well-being from large-scale industrial farming and product distribution, but some were skeptical and cautious about the trade-offs required to support such segregated food-production systems. Even Olmsted questioned the future and value of nineteenth-century urbanization after witnessing wholesale destruction and radical alteration of natural landscapes to make way for ever-expanding population centers. He warned if functional green spaces were not maintained or incorporated as part of urban design, "the employment of simple and sensible social life in our community seems likely to be entirely destroyed."[5] By the end of the century, such thinking manifested in the form of community gardens, first in Detroit, followed by New York City and Philadelphia.

Struggle—1900s

The community gardening program on vacant lots in Detroit at the end of the nineteenth century focused on alleviating immigrant hunger and poverty caused by economic hardship. The program, which supplied instructions, support, and tools to establish garden plots, was highly successful and replicated across the country. However, many programs declined when the economy improved. Beginning in 1898, the Garden City movement exemplified a formal program for enhancing food production in built areas. The movement persisted into the early 1900s, with most projects focused on social reform. There is some evidence that the movement tied together the poor and middle class by integrating access to affordable housing. The Garden City movement was based on an urban planning method developed by Sir Ebenezer Howard of the United Kingdom, which aimed to create self-sufficient cities that were proportionately divided into residential, industrial, and agricultural areas, all of which were surrounded by a continuous natural area called a greenbelt. His idea was that the entire estate—both town and agricultural sectors—would be held in common ownership.

According to Howard, the population of such a Garden City should be 30,000 people on an estate of 6,000 acres, or a size that makes possible a full measure of social life, but no larger. The physical limit on the expansion of the city was the greenbelt, which would be permanently zoned as agricultural and occupy five-sixths of the entire Garden City. It was also meant to serve an economic purpose by integrating agriculture into the urban system, thereby revitalizing the rural economy. Howard also promoted the use of railways and roads to connect large cities and nearby smaller population centers called satellite cities. The idea spawned many developments in the United States in the early twentieth century, such as Greenbelt, Maryland; Greendale, Wisconsin; and Greenhills, Ohio; among others. The main premise behind Howard's approach was to combine the best of both urban and rural life.

Howard's proposal was overly prescriptive and technocratic, but it represented an emerging mode of thinking in contrast to the segregation of urban and rural land use. It inspired a movement intent on envisioning new development and built working environments that made prominent space for greenways and agriculture in urban areas. As perspectives on including green space for recreation and food production grew, so too did the American viewpoint regarding the management of forests. The first forestry school in the United States was established by Carl A. Schenck near Asheville, North Carolina, on the Biltmore estate owned by the Vanderbilt family. The large swath of forest where Schenk taught the first generation of foresters how to manage a forest scientifically without clear-cutting is now part of the Pisgah National Forest and known as the Cradle of Forestry. The foundations for understanding forestry, ecology, and urban gardening were being established. Around the same time, Franklin Hiram King, an American agronomist, introduced information on organic farming learned while exploring and observing farming traditions in Japan, China, and Korea.

Tragedy and Advancement

During the second decade of the twentieth century, two forebearers of agroecosystems and food forests were born in Japan and England: Masanobu Fukuoka and Robert A. Hart. Both would go on to write influential texts: *One Straw Revolution* and *Forest Gardening,* respectively.

During the First World War, the federal government promoted school and community gardens to increase domestic food supply so more could be shipped to soldiers. Agricultural education was incorporated into the school curriculum through a program called the United States School Garden Army. Biodynamic farming, which is a mix of organic agriculture and concepts that view farms as a single organismic system, began to be practiced in Europe. Systems thinking was also emerging in fields such as organismic biology, Gestalt psychology, and the advancing science of ecology.

When the Great Depression hit, many people lost their farms as agricultural prices dropped. The tragedy prompted the push for a national farm bill and its government subsidies, which had a significant effect on agricultural practices. There was also a resurgence in gardening, primarily driven by wartime austerity. For instance, more than twenty million Victory Gardens were planted during the Second World War, which supplemented the food supply for many families. In addition to augmenting rations during wartime, Victory Gardens yet again linked production in urban areas to values associated with overcoming social and economic strife and national pride during crises. This reinforced the pattern that once national obstacles were overcome, gardens were abandoned as a way of putting tough times behind and symbolizing forward progress. In their place, a return to modern convenience was encouraged to demonstrate headway.

Another icon of advancement was the automobile, which forever changed the character of American society. Discretionary, fast, and mostly inexpensive transportation became accessible to the general public, and a profound shift in settlement patterns followed. With their automobiles leading the way, Americans began moving to outlying suburbs. Suburban growth was fed by both city dwellers seeking home life at some distance from city hustle and bustle and rural people who wanted to be closer to amenities. One result was increased land conversion around cities to make room for sprawling developments with manicured lawns. By the time of this social change, food production had ceased in all but a few remnant places in most city centers.[6]

Expansion and Industrialization

Paradoxically, while organic agricultural practices were being formalized and Jerome Rodale, an early thought leader in the organic movement, began publishing *Organic Farming and Gardening* magazine, the Green Revolution was simultaneously ramping up to feed the world. The juxtaposition of these two forms of agriculture helps frame how far the country had split on agricultural practices and food preferences. Food culture had transformed over the course of 200 years from distributed subsistence based mostly on localized hunting, gathering, foraging, and small cultivation to one of wide-ranging convenience separately sourced by large-scale remote and compartmentalized agriculture and animal husbandry operations. The change was made possible by advances in mechanization and production science, both of which ensured majority ownership of competitive production among powerful industrial interests.

Increasingly sourced by production systems with massive economies of scale and shaped by assembly-line labor, America's food culture largely became one of expectation and preference for ease and assortment. Supermarket convenience freed people from the hard labor of farming and gardening. Easy access to food helped create a highly efficient workforce and fostered an unprecedented period of development. Ample availability fueled a rise in the quality of life for many Americans. A remarkably flexible life took root by the mid-twentieth century in America, and a new interest in nature surfaced with activities such as rural recreation and landscape aesthetics defining much of the focus.

The construction of the interstate highway system and mass expansion of roadways in the 1950s increased suburban sprawl and helped precipitate decline in many city centers. Urban landscape decay would continue for several decades, with public attention turning largely to externalized environmental problems resulting from a hundred years of rapid industrialization. But this concern was not necessarily focused on neighborhoods, suburbs, and other built environments. Rather, it was often aimed at a separate nature that was far away and whose purity was revered and to be protected in the name of public conscience and to afford passive consumption such as recreation. The stance took on a similar resolve as that of the sentiments of rural values and nobility espoused by the pleasure grounds experience of the nineteenth century, but was more holistic and ecological. The concern for protecting nature was driven by sizable change in American cultural values and awareness toward the environment in the 1960s.

Environmental Awakening

Following the chestnut blight that eliminated entire populations of a once abundant species, Dutch elm disease devastated urban forests where American elm was a popular choice for street trees. Suddenly many city streets were left bare. Nature's fragility was witnessed on multiple accounts while social resilience was being tested through counterculture activism spanning antipoverty measures, food production and distribution, the Vietnam War, and the civil rights movement. Demonstrations were in full swing on these issues, and others arose to protest the exploitation of natural resources and lack of environmental regulations.

The 1970s was a period of environmental awakening. The Environmental Protection Agency was established, policies such as the Clean Air Act were passed, and the first Earth Day was celebrated. Nested in this environmental enlightenment was growing concern about food quality and safety. The first organic farmer certifications surfaced, and community gardening was rebirthed in response to urban abandonment, rising inflation, and environmental concerns.[7] The late 1970s witnessed a formative effort toward development of the local food and sustainable agriculture movements that are common today.

The "baby boom" of the 1950s vastly transformed the nation's social consciousness for years to come, the composition of which was fueled by sheer numbers but also by the interplay between a new postmaterial American spirit and the artifact of manifest destiny articulated some 150 years earlier. Initially intended to convince a young country of its physical independence and purpose, manifest destiny resulted in a romantic westward exuberance combined with the intellectual stirrings of an educated class. This set the stage for a profound period of social awareness and activism. It was during this time that a modern-day environmental class emerged. Inspired by works such as Rachel Carson's *Silent Spring* and empowered by federal policies that elevated public participation in land-use decision making, this period fostered a reemergence of urban greening and food production. It also set the stage for the surge in community food forests we are now experiencing in the twenty-first century.

CHAPTER 7

The Dr. George Washington Carver Edible Park

When studying community food forests for this book, we noticed that many teams describe their projects in terms of food literacy and accessibility, community health and well-being, and environmental sustainability. We also found that community change is part of the conversation, but unlike food forest design, details about exactly how leaders will do this and what they intend to change are often lacking. In the end, goals such as changing local food systems by installing a community food forest are immensely complex; they take time and involve multiple stakeholders with different viewpoints and responsibilities. Nevertheless, permanence and change are possible, but perseverance is necessary and a little serendipity helps, too. We know of no better example of this than what transpired at the community food forests in Asheville, North Carolina, over the past two decades.

As we discuss later in this book, critically reflecting on food accessibility and nutrition among all community members is paramount when working toward change, particularly through community food forests. Successful teams are those that roll up their sleeves and get to work on the hard tasks of uniting organizations and individuals working on similar causes. Though the Dr. George Washington Carver Edible Park in Asheville was not immune to potentially fatal problems such as turnover in leadership, questions and concerns among public representatives, too much development too fast, and times of limited to no maintenance or use, its continued existence is a testament to the power of perennial food forest systems. When humans needed to step aside and sort out their own complexities and processes, the food forest in Asheville continued to grow and remained ready for the next step. By its very existence, in good times and bad, the community food forest contributed to change by offering a place where like-minded people could come together and push food accessibility and environmental health into the public eye.

Fast Facts

PROPERTY OWNER: **City of Asheville Parks and Recreation**
LOCATION: **Stephens-Lee Recreation Center, 30 George Washington Carver Ave, Asheville, NC 28801**
YEAR PLANNING STARTED: **1997**
YEAR ESTABLISHED: **1998**
SIZE: **Currently 1.1 acres, started as 1.5**

ABOUT ASHEVILLE

Asheville is nestled in the Blue Ridge Mountains of western North Carolina and inhabited by about 87,500 residents. The Cherokee once paddled trade routes through the region on the Swannanoa and French Broad Rivers, which early European explorers such as Hernando de Soto followed during westward expeditions. More than two hundred years later, Asheville is known for its creative class, historical architecture, mountainous landscape, and diverse and vibrant culinary scene. Tourism is one of the city's largest industries, and more than nine million people visit each year to dine, socialize, and see attractions such as the Biltmore Estate, Grove Park Inn, and Blue Ridge Parkway.

In 2007, Asheville established itself as a culinary destination by claiming to be the world's only Foodtopian Society. By 2012, the Buncombe County Tourism Development Authority had trademarked the Foodtopia and Foodtopian Society brands to build exclusive content in support of the local restaurant scene (Asheville is located in Buncombe County). The city created slogans, logos, blogs, and a website that pushed foodtopian mantras such as "we believe that a community's quality of life is directly proportional to the quality of its food" and a "community of culinary collaborators crafting an experience to nurture your soul."[1] In 2011, right around the time Foodtopian branding took off, the Food Research and Action Center[2] ranked Asheville number seven in the country for food hardship;[3] in 2012, it jumped to third place.[4]

To confront Asheville's growing food disparity, the Asheville Buncombe Food Policy Council (ABFPC) formed in 2011. Their efforts and those of many others began to pay off, and in January 2013 the city of Asheville approved a food policy goals and action plan drafted by the ABFPC.[5] These goals were subsequently incorporated into the city's sustainability management plan. In 2017, the goals were updated to reflect the city's efforts to make parks, public lands, and greenways a regional model for public edibles to increase food access and distribution. Additionally, the parks and recreation department agreed in the new action plan to expand annual and perennial plantings through partnerships, maintenance, and communication.

IDEA GENERATION

Jonathan Brown was a college student in the mid-1990s when he first heard about food forests. He was attending Appalachian State University in Boone, North Carolina (near Asheville), where he was minoring in sustainable development. While there, he learned about permaculture, which sparked his interest so keenly that he later completed two intensive design classes and an instructor's course, eventually becoming a certified permaculture educator. He met Samantha Lefko, a fellow Appalachian State student who self-designed her major around the impacts of public education on social structure.

When Jonathan gave a talk on campus about permaculture, Samantha became interested in its potential applications. After spending time studying abroad, she returned to the United States and obtained a certificate in permaculture design. She returned to Appalachian State and reconnected with Jonathan through coursework on sustainable development. The classes focused on developing countries, but Samantha and Jonathan shared an interest in transferring what they were studying to communities in the United

THE DR. GEORGE WASHINGTON CARVER EDIBLE PARK

FIGURE 7.1. The Bountiful City Project was officially renamed in 2000 to honor Dr. George Washington Carver, a scientist, botanist, educator, and inventor (1864-1943) whose life story and legacy of agricultural work are told in the book *The Man Who Talks with the Flowers*. The book inspired the entrance sign depicting a silhouette of Dr. Carver holding a flower. The sign is a conversation starter, especially on tours of the site, demonstrating how symbols and art can be used in food forests to enhance cultural significance.

States. They believed cities would be the future and should be a focus of sustainability initiatives.

Samantha and Jonathan married and decided to move to Asheville after graduation. They came upon a handful of community garden programs in the city, but none that were completely open to the public. The original goal for many of these projects was to turn vacant lots into community gardens for use by apartment dwellers or others who lacked garden space. Samantha and Jonathan set out to build upon the existing community garden network, but they aimed to do so using self-regulating perennial systems that would provide food with less maintenance and thus increase possibilities for broader public use. They wanted to create a new type of community garden: a public place and edible space where anyone could enter and freely pick an apple, pear, or other available food at any time of the year.

Most of what Samantha and Jonathan had learned through permaculture courses was geared for small farms or backyards. They received little instruction on how to complete a community-level project. Because of this, they decided to pursue a hybrid model that borrowed from both permaculture and sustainable development philosophies to create the edible park concept and planning process. They were able to integrate sustainable development techniques such as public charrettes for gathering community opinion. Another strategy was to involve the public in every step of planning, planting, and maintaining the project to develop community ownership for long-term sustainability.

Samantha and Jonathan formed a nonprofit organization called City Seeds, and they quickly fleshed out many ideas for projects, one of which was an edible park initiative they called the Bountiful City Project (which was later renamed the Dr. George Washington Carver Edible Park). The goal was to combine elements of parks, community gardens, and permaculture to create edible public spaces in urban and suburban areas. The Bountiful City Project was innovative not because of what it would accomplish, but because of where it would work and who would be involved. It sought to retrofit public land and ensure full public accessibility through an innovative effort that would intentionally combine the concept of public commons with contemporary perennial landscape design inspired by permaculture. The collective processes used in sustainable development would be applied to work across diverse groups of civic organizations and public agencies.

BUILDING SUPPORT AND FORMING PARTNERSHIPS

City Seeds worked hard behind the scenes before publicizing and promoting the Bountiful City Project. They pursued funding, acquired land, assembled partners, developed plant palettes,

The Name Game

Samantha and Jonathan initially chose Edible Entropy as the name for their organization, but quickly learned an important lesson: A cryptic or fancy name can become an obstacle to progress. They soon changed the name to City Seeds—the simpler name meant less time explaining the organization's name and more time discussing its potential. Although the organization they created is now defunct, other organizations have appeared with the name City Seeds, such as one in Baltimore, which is a culinary social enterprise. A good name may indeed become popularized and used elsewhere, but most important is how it helps attract people. A good name is easy to remember and associate with organizational goals.

and designed the food forest. The organization was successful in obtaining funding from foundations because its goals were novel and funders believed they showed great potential. For example, they received a $10,000 grant from the Community Foundation of Western North Carolina and $15,000 from the Z. Smith Reynolds Foundation, among others, to begin.

City Seeds began bidding on land, but nothing affordable panned out. Donated land emerged as the only realistic option, which led to pitching the project to Asheville's parks and recreation department. To prepare for making the case, City Seeds analyzed how the project aligned with the city's comprehensive plan and its focus on sustainability and food production. City Seeds argued that the project would be a useful step in helping Asheville become a green community, as the city had defined itself.

The first meeting between City Seeds and Asheville included Al Kopf, the city's parks, planning, and development superintendent. Al was a landscape architect with training in archeology. He had served as an agricultural Peace Corps volunteer in Ecuador where he helped with reforestation of underutilized farmland to offset deforestation and protect soil. He and other staff members were excited by the Bountiful City Project. Al appreciated the parallels between the work he had done in Ecuador and the proposed retrofit of idle city property as an edible park using perennial species. In March 1997, City Seeds signed a cosponsorship agreement with the Asheville city manager that outlined the roles of each partner. Al Kopf would serve as the city's liaison to help move the Bountiful City Project from concept to reality.

Samantha and Jonathan visited several potential sites, but none of them seemed like a good fit until they toured a one-acre lot at the Stephens-Lee Recreation Center. Ecologically speaking, the site had problems. Aggressive kudzu vines had taken over along with grasses and invasive plant species. Soil tests revealed, though, that the site was highly suitable for a food forest. Nitrogen needed to be added, but City Seeds considered this an opportunity to showcase the positive benefits of a permaculture approach. The food forest design would include plants that naturally make nitrogen more accessible. Since the site needed a lot of work, using it for the food forest was a win-win. The City of Asheville did not have sufficient staff or finances to improve the land without help from City Seeds. City Seeds would benefit from

Preserving a Project's Story

In 1997, local news stories did not go viral on the internet and social media as we now know it. The Dr. George Washington Carver Edible Park made local news but did not receive national coverage. Likewise, little was archived online about the project. A decade later, when community food forests began emerging in numerous locations, few project leaders would know of earlier work in Asheville. Luckily, newsletters, articles, reports, and work orders were archived through a partnership with the city's parks and recreation department. This store of information helped document how the community food forests evolved in Asheville.

Creating social media sites and building online repositories of information about a project enhances public access and knowledge about community food forests in general, while also providing specific examples from which to draw for inspiration and insight. Archiving information in any form helps people pick up and move forward in the event there is a lapse in organizational or community support.

the in-kind contributions from the parks and recreation department such as bulldozers, tools, municipal compost, and materials for a boardwalk and signs. The city also dedicated employee time to help plan and install the park.

Al knew that the site where the food forest would be located was once the grounds of Stephens-Lee High School, a central part of

FIGURE 7.2. In 1997, the site for the Dr. George Washington Carver Edible Park was mostly grass and invasive species. Volunteers worked hard to clear it and prep the soil so they could install trees, plants, and a walkway. The site was densely vegetated by 2014. Photos courtesy of Jonathan Brown.

Asheville's historic black community. Stephens-Lee High School was the only secondary school for African-Americans in western North Carolina from 1922 to 1965 and was known as "the Castle on the Hill" due to its beautiful architecture and location overlooking the neighborhood. In 1975, the school was closed following desegregation, and most of the building was demolished. However, the gym was left standing and is now used as the Stephens-Lee Recreation Center. The park would be located down the hill from this building, on the spot where the main school building once stood. Soil had been excavated from the site in the years following the demolition to be used elsewhere in the city.

It was believed that locating the edible park at the site of the old high school in one of Asheville's most important social landscapes would help renew valued community space, provide access to healthy food, and serve as an important educational venue for youth. The park was envisioned as a test site, an experiment with great promise and profound possibilities. If successful, another food forest would follow at another parks and recreation site located in the historic neighborhood of Montford.

The Relevance of History

Understanding the historical context of a neighborhood or city section where a community food forest will be placed can be an asset. For example, history can be used to imbue meaning in a site or help to bridge cultural divides during the planning phase. In addition to the specific history of the site where the Dr. George Washington Carver Edible Park stands, the city of Asheville has cultural history that is relevant. The Biltmore Estate, completed in 1889, remains the largest single-family residence in the United States. It was built by George Vanderbilt, who also bought 125,000 acres around the estate. While many visit the Vanderbilt Estate annually, not many know that much of the land was acquired by buying out African-American farmers who located there at the end of the Civil War. They were moved off the property, along with their church and cemetery, to a new location about two miles away.[6] This community, which still exists, is known as Shiloh.[7]

The Shiloh community formed a nonprofit association in 2000 to unify residents and build a stronger voice in local politics in the face of ever-expanding development. In 2003, the association paired with the nonprofit Bountiful Cities and the Asheville parks and recreation department to establish and maintain a community garden with fruit trees open for harvest to community members within its neighborhood. Relocating the Shiloh community and demolishing the Stephens-Lee High School were disruptive events in Asheville's past with lasting impact on members of the community. The community garden and orchard created in the Shiloh neighborhood was a constructive and regenerative project that has positively served some of those impacted community members in multiple ways. For example, the pavilion located at the garden provides shelter for the annual historic Shiloh Community Celebration when people gather for food, friendship, and stories of Shiloh.

The food forest created by City Seeds on the old grounds of the Stephens-Lee High School was also turned into a constructive project that recognized the past. In 2000, Lady Gloria Howard Free, an alumna of Stephens-Lee High School, a board member of City Seeds, and a prominent leader in Asheville's black community, suggested changing the name of the food forest from the Bountiful City Project to the Dr. George Washington Carver Edible Park. That same year, the nonprofit Bountiful Cities formed to help sustain momentum and coordinate similar projects in and around the city.

It was a step in using the community food forest to draw attention to and symbolize pride in an exceptional academic and researcher who is known for his contributions to alternative crop production and soil conservation. A large sign was posted at the entrance to the community food forest from the Stephens-Lee Recreation Center parking lot to enhance cultural connections to the site.

Discussions about park planning and implementation continued between Asheville and City Seeds. In the meantime, Jonathan and Samantha researched topics related to sustainable urban design and food systems. They looked all over the country for examples that they could replicate in the park's design, but they found nothing. The plan called for the park to be installed in phases. At each step, the focus would be on one vegetation layer accompanied by an educational workshop and workday with community members. There were plans for additional workshops, too, such as growing and using culinary and medicinal herbs, dehydrating fruits and vegetables, improving soil, and saving seeds. City Seeds also began partnering with groups such as the YMCA and the Stephens-Lee Recreation Center to develop workshops with a youth focus.

An important selling point for the project was that site maintenance would decline in three to five years, but food yields would continue and serve as a place for harvest-day celebrations, food and environmental education, and ongoing community activities. To evaluate the project's success, several metrics were proposed:

- Keep annual records to estimate production and assign dollar values.
- Track workshop attendance.
- Document interest among local organizations.
- Account for the number of food forests developed by workshop participants on their own property.

This last piece would be particularly important because a key goal of the project was to spark establishment of private food forests. City Seeds was ready to assist those who were interested.

City Seeds put together a board of trustees to help develop the Bountiful City Project vision. The board included public and private members such as a planner for the city of Richmond, Virginia, a local permaculturist and teacher, the former greenway planner for the city of Asheville, and professors from the University of North Carolina. City Seeds and the city of Asheville also wanted to involve Asheville residents in project development in order to ensure buy-in and sustainable use as well as provide meaningful outreach and communication.

INSTALLATION AND COMMUNITY ENGAGEMENT

In 1998, Asheville opened a period of public comment to hear citizen perspectives on the project's purpose, philosophy, and anticipated benefits. Partners in the Bountiful City Project also worked to garner support from the Stephens-Lee Alumni Association and the surrounding community and nearby church by connecting with community leaders. In collaboration with the parks and recreation department, City Seeds planned to gather information for the initial design by hosting a public design charrette meeting at city hall. Flyers inviting the public to help design their "dream park" were distributed in the community and word-of-mouth invitations were encouraged through various channels including local health food stores and restaurants, City Seeds' newsletter, community groups, and the local newspaper.

A design charrette occurred in early 1998 and consisted of presentations on the food forest concept, how such systems might look and function in Asheville, and an overview of the preliminary project idea including pictures of the Stephens-Lee Recreation Center site.

Around twenty participants were asked to form small groups and then given lists of suggested and necessary elements to consider in their design. Next, everyone visited the site, which was within walking distance of city hall, to take notes on characteristics such as topography, water, and sun/shade patterns. They were also asked to consider the following questions:

- What should be the center of the site?
- Should there be a centerpiece?
- Where would entranceways be best placed?
- How many pathways should there be and how will they flow?
- Where are the best views?

Charrette results were shared with participants two weeks later to demonstrate open communication and maintain momentum. City Seeds modified earlier designs based on the charrette suggestions and was ready to move forward with the first public planting event at the site on March 21. In anticipation of the event, the city provided large supplies of mulch and used its heavy equipment and bulldozers to prepare the site. Shortly after the small inaugural planting, a cypress boardwalk was installed to allow access to the site even when stormwater was present. The boardwalk also served to protect newly planted culinary and medicinal herbs from being trampled. A year later, City Seeds extended the boardwalk and constructed a small stage lined with blueberry bushes to create an edible amphitheater with berm seating.

Planting and installation continued through the summer of 1998. A second public design charrette occurred in late September. The exercise included an overview of phase one accomplishments and presented concepts of public space aesthetics associated with community gathering

Developing Metrics

Bountiful City Project's main purpose was to provide hands-on educational program and learning. Programs give participants (particularly children) lifelong gardening skills, help instill respect for the environment, and develop appreciation for the neighborhood and city. City Seeds believed nurturing an edible park was the first step in developing patience, concentration skills, and caring for self and community. However, they needed to develop evaluation strategies and metrics to gather evidence about project impact. Developing metrics for tracking the various benefits and impacts public food forests have on participants and communities is difficult. Measuring harvestable yield in the form of produce is a tangible yet limited metric, because many benefits of a community food forest are social, psychological, and physiological. We recommend thinking about metrics from the beginning that will help you understand the effects of your project and account for educational impacts. Questionnaires can be developed and distributed to participants before a workshop or at the beginning of the growing season to evaluate their knowledge on a particular subject or range of gardening skills. A follow-up questionnaire at the end of the workshop or growing season is one way to track results and account for change. It can also show gaps in programs if desired skills are not increased and can help drive planning to improve outcomes in the future.

FIGURE 7.3. Blueberries need acidic soil conditions to grow best, which often requires amending the soil when including them in a food forest. These blueberry bushes are near the entrance of the Dr. George Washington Carver Edible Park, with easy access to a nearby parking lot, which made it possible to bring soil amendments to the planting site. Planting highly recognizable fruit near a food forest entrance is a good strategy to draw attention and interest.

places. Participants were again invited to form small groups and asked to visit the site and make their own assessment of progress. They then returned to city hall to sketch out and present ideas for next steps. Evaluation forms created by City Seeds were available for participants to fill out to provide feedback on the process and how to improve it in the future.

The following month, City Seeds hatched a strategic partnership with Earth Fare, a well-known local health food store, to be the business's charity of the month. This opportunity allowed City Seeds to maintain a table outside the store during weekends to build organizational awareness, solicit donations, and inform the public about the Bountiful City Project. The park's all-day grand opening at the Stephens-Lee Recreation Center in October 1998 included support from various organizations, businesses, and agencies in the Asheville area who provided materials and financial support. The event included storytelling, tours, plant identification training, and workshops on pond establishment and small-scale food production. More than 150 people attended.

Discussions began in November 1998 on developing a second community food forest in the historic Montford neighborhood of Asheville, and a site at the Montford Recreation Center was soon chosen. In January 1999, the public charrette process began again. By this time, word had spread about the Bountiful City Project and approximately fifty people attended the charrette. City Seeds signed another agreement with parks and recreation to create a community food forest in Montford and dedicated the remainder of the year to developing site design, applying for grants, and organizing a small conference in Asheville on urban ecology. Their efforts paid off and they received a $5,000 environmental education grant from the Environmental Protection Agency by the end of the year.

By January 2000, the estimated budget for the new site was approximately $8,000. However, mixed feelings were developing about the pace of development. The city was considering a $500 mini-grant to purchase plant material for

THE DR. GEORGE WASHINGTON CARVER EDIBLE PARK

FIGURE 7.4. People discover community food forests in surprising and varied ways, including this couple on the boardwalk at the Dr. George Washington Carver Edible Park in Asheville, who first learned about the site through a geocaching exercise. In the photo on the left, notice the dense forested tunnel that leads into a patch of light. Food forests will require canopy management to achieve preferable light conditions as it matures. The photo on the right shows an area of the food forest where trees have been pruned over the years to maintain acceptable levels of light along a walking path and bridge that connect downtown Asheville and the neighborhood where the food forest is located.

the new site. Yet, at the same time, some were concerned that progress was too rapid and key management considerations were being overlooked. To address this, the parks and recreation department took steps to establish an agreement outlining the responsibility of each partner in terms of site maintenance and structural quality, standards of use, and program monitoring and evaluation for meaningful public service.

CONTINUATION AND FUTURE VISIONS

The cautionary feelings about the speed at which the new site was being developed were perhaps justified. In 2000, Samantha and Jonathan left Asheville to pursue higher education, and City Seeds ceased to exist. The edible parks were left without central leadership and care and maintenance started to decline. Around the same time, a new urban agriculture organization led by Darcel Eddins formed and took on the name Bountiful Cities to keep the energy going, but it was a separate, stand-alone nonprofit organization. Bountiful Cities first focused on maintaining a community garden, the Pearson Garden, in Montford. By 2002, Bountiful Cities had signed an agreement with the parks and recreation department to assume responsibility for the Dr. George Washington Carver Edible Park to include coordinating volunteer workdays and educational programs.

The Buncombe Fruit and Nut Club began helping Bountiful Cities by pruning trees and providing general upkeep at both the Dr. George Washington Carver Edible Park and the Montford site. This local club prided themselves on being a nonhierarchical group that plants and maintains fruit and nut trees on public property. The Fruit and Nut Club has a slightly different approach to their publicly available harvest. Rather than planting everything in one location, they planted trees in low densities in many areas of a municipality. Although anyone can find one of the trees and freely harvest, only members of the group that donate volunteer time gain access to a map that shows the locations of club-planted trees throughout the city.

Another partner was Asheville GreenWorks, an organization with a mission centered on stewardship, advocacy, beautification, and education for the natural resources of Buncombe County. GreenWorks was a perfect partner with which to launch a community food tree project. The goal was to plant orchards throughout Asheville, starting with low-income communities. Both organizations advocated that projects should abide by the terms of community members, hold stakeholder meetings and workdays when and where residents prefer, and offer multiple ways for the public to provide input and participate.

Although three organizations plus a host of local individuals were interested in the Dr. George Washington Carver Edible Park, Bountiful Cities maintained overall stewardship responsibilities. They were tasked with keeping the city of Asheville informed on progress and accomplishments. They were also required to manage a formal process of volunteer registration and tracking for liability and insurance purposes. Like most nonprofits, Bountiful Cities was involved in many projects and was mostly run by volunteers. Coordinating partnerships, volunteer workdays, and educational opportunities is a lot of work; Bountiful Cities's main focus was on community gardens in low-income neighborhoods, so the food forest was not necessarily a top priority. As the site matured, the canopy filled in and created a heavily shaded understory where few low-growing edible plants held ground. This left openings for invasive exotics like kudzu, resulting in a continuous need for maintenance. A Bountiful Cities volunteer coordinator, who managed the site for five years, realized it was time to revamp management strategies and circle back to the community to create a new shared vision.

Bountiful Cities organized a stakeholder planning meeting in 2015 to discuss next phases of management, renew use agreements, and reaffirm terms of partner maintenance and responsibilities. The goal was to identify a way forward that would help renew interest and strengthen capacity in the community food forests. The outcome was six vision statements for the Carver and Montford community food forests regarding safety, education, agroforestry, infrastructure, community, and cooperation. Agroecological goals included:

- Creating a genetic seed bank
- Using on-site cuttings to propagate desirable species
- Scionwood harvesting for grafting and marcotting (a form of propagation also known as air layering)
- Promoting regionally adapted plants available for the public to propagate at home
- Encouraging use of native plants for pollinators

Enhancing understory vegetation using edible herbs and flowers on raised beds was also suggested to help reduce maintenance needs for undesirable plants. Another important suggestion was adding more features such as trails, lighting, a butterfly garden, places to picnic, and trash cans. Other goals pertained specifically to education:

Safety and Maintenance Considerations for a Shaded Understory

Even though it is tempting to plant trees close together when they are small, the Dr. George Washington Carver Edible Park demonstrates how trees planted too closely together can become crowded after several years. Heavy pruning or even removing large vegetation is regularly needed to allow proper air flow around fruit trees and to create patches of sunlight for understory species. Increasing the amount of light in the understory is important from ecological and social perspectives. Although the shaded site makes the park feel like a secret garden for some, it also blocks line of sight in several locations, making it feel unsafe for others. At the Carver Edible Park, there is an obscured empty field nearby that is hidden from plain view by large evergreen trees between it and the road, which is known to be a favorite camping spot of transient visitors. The path that cuts through the food forest connects the neighborhood to downtown Asheville via a well-used walking bridge. In 2013, community members began expressing concerns about safety while passing through the densely forested area, particularly at night. During a stakeholder meeting, police monitoring was discussed along with eliminating potential hiding places to ensure a safe and inviting space suitable for people of all ages. Eventually some fruit trees were removed to make room for park lighting.

- Creating an agroecology learning center with programs for all ages
- Holding a wide range of hands-on events such as workshops, tours, and removal of invasive species
- Holding classes after school and summer camp programs for youth
- Installing signs at both entrances
- Installing informative trail placards and a site map
- Creating a plant identification guide

Several programs aimed at reviving community interest were offered, such as yoga in the park, quarterly socials, novel educational programs, and community events that connected to cultural heritage and created a safe space for people to gather and socialize. Stakeholders thought it would help to work more closely with neighborhood residents to direct activities, plan programs, maintain trees, and harvest food. Increasing the amount of family activities was an important goal, and creating a strong partnership with the Stephens-Lee Recreation Center was seen as a way to increase buy-in. Overall, the group agreed that cooperation and clear communication could help maintain ongoing support, generate greater publicity, and strengthen networks with other neighborhood organizations. The hope is that additional resources will be secured in the future to fund a position for a park steward.

In fall 2017, a public participation program was held to help design lighting for the edible park. The local neighborhood association, the new director of the Stephens-Lee Recreation

Center, a parks and recreation department program manager, the director of Bountiful Cities, and neighborhood residents walked the site at night with the landscape architect responsible for creating the lighting design. As they walked along the path, they stopped periodically to discuss areas of concern or with potential for lighting. They talked about how the lighting would affect the food forest or where vegetation could be altered to improve visibility. In the end, everyone had an opportunity to be heard, which helped avoid misunderstanding or miscommunication and strengthened collaboration among all stakeholders. A renewal is also underway for 2018 in terms of youth programming, with efforts focused on reinstituting planting projects, educational tours, and a harvest celebration.

PART 3
Functional Design

CHAPTER 8

The Role of Agroecology

Agroecology applies ecological principles and practices to agriculture in order to create sustainable and resilient food production systems. The field has played an important role in shaping community food forests in the United States. Agroecology's principle of eco-mimicry is evident in community food forests because a forest ecosystem is used as the template for design. On the other hand, ethnobotany, which is used regularly in the field of agroecology, is less visible in physical design but appears in the mission statements of many organizations that use community food forests to reconnect people and plants. In short, agroecology focuses on harnessing nature's energy instead of struggling against it when producing food, fiber, and other benefits from the land.

Agroecological practices ultimately provide a pathway for achieving environmentally sustainable food production in a community food forest. Project leaders pursue the goal of working with, and learning from, nature. Environmentalism, diverse production, local economic stability, and food literacy and security are key principles of agroecology, and much of how we design and manage community food forests reflects these principles. People strive to make community food forests rich agroecological systems with useful species that provide benefits to a community that will enrich people for years to come.

Agroforestry and permaculture are well-recognized agroecological practices. Most community food forests use elements of both. Agroforestry is a set of scientific practices reflecting traditional ecological knowledge. Trees that offer specific economic and environmental benefits are managed in combination with agricultural crops and/or livestock. Medicinal herbs and edible mushrooms are also farmed in woodlands. Permaculture, on the other hand, is a process that uses particular techniques that emulate patterns found in nature. It is largely based on observation and a sustainable land use philosophy rather than scientific principles. It is also a worldview that seeks to build sustainable agricultural systems, environments, and communities ethically. A key goal of permaculture is to promote land use systems that are self-generating and managed adaptively to sustain production over time.

In many ways, agroforestry and permaculture resemble each other. Rather than managing for a single crop (a monoculture), both combine an array of crops (a polyculture) and include a mixture of perennial and annual species. Plants and trees in these polycultures are sometimes planted in rows or rings; other times, they are randomly clustered. However, plantings are always intentional and carried out with environmental goals in mind. When designing an agroforestry or permaculture project, one of the most difficult tasks is determining optimal levels of species diversity and density. These important decisions affect growth and production levels and require research and careful planning.

Community food forests also include these complexities. What makes them different from agroforestry on a farm or a permaculture project in a backyard is that they include a human community. Success is not simply a matter of how well plants and trees work together in the same space. It also depends on the extent to which the public is culturally engaged, physically involved, and intellectually inspired. However, deliberately connecting projects and community by intentionally using agroecological practices can help achieve goals of sustainable food production, environmental benefits, and land rehabilitation.

AGROFORESTRY IN COMMUNITY FOOD FORESTS

Agroforestry combines land management science, indigenous knowledge, and traditional farming practices. Early forms of traditional ecological knowledge among North American indigenous tribes encompassed agroforestry-like polycultures. Forests were managed for food and material resources such as acorns, chestnuts, and pine nuts; other varieties of canopy species; forest animals; tree saps; and understory species such as berries, grasses, herbs, and mushrooms. In the early twentieth century, scientists studied indigenous forms of multicropping to develop a verifiable and reproducible system based on age-old practices. The researchers looked at tribal practices and rural family farms worldwide. They drew from practices such as taungya intercropping in Burma, Mayan home gardens, Indonesian multistory cropping systems, and North American windbreak plantings after the Dust Bowl in the Midwest. It was the beginning of a concerted effort to develop a discipline, with substantial strides made in the 1970s and early 1980s.

Agroforestry in temperate climates is not as well understood as practices developed in tropical and subtropical climes. Yet it is gaining ground, particularly as farmers seek alternative planting methods to adapt to climate change. Today, there are a mix of temperate agroforestry practices that producers in the United States can use. These practices are carried out on a small number of rural farms and homesteads. Often, these producers are using practices that fit under a broad definition of agroforestry but may not fit neatly into the five practices historically recognized in temperate regions: windbreaks, riparian buffers, forest farming, silvopasture, and alley cropping. These practices are even less common in suburban and urban settings. However, this is changing as agencies such as the USDA begin adding urban agroforestry practices to the list of applications they support through education, outreach, and grant partnerships.

Tropical home gardens are perhaps one of the oldest agroforestry practices. They are typically multilayered mixed forest gardens with a wide range of species that are common in places such as Indonesia, India, and Central America. These traditional practices generally differ from home gardens in the backyards of places like the United States, which are generally more systematic and often designed using three broad structural layers, ranging vertically from ground cover through the subcanopy to canopy. Within each

THE ROLE OF AGROECOLOGY

FIGURE 8.1. Food forests serve as multifunctional riparian buffers (an agroforestry practice) when placed alongside waterways in urban areas. This newly planted community food forest in Greenbelt, Maryland, received funding through a project to protect waterways leading to the Chesapeake Bay. *Above*, the food forest has been mulched and planted with young edibles near a playground. *Below*, a minimal vegetative strip buffers the stream. The outline at the back of the photo indicates where the food forest extends the forest edge and enhances the width of buffering vegetation.

layer are many sublevels that are filled by a variety of plants. In the temperate agroforestry domain, this is sometimes referred to as multistory cropping. Agroforestry acknowledges that categorizing all configurations within set standards of forest garden design is practically impossible due to the variety of possible plant combinations. The spatial arrangement and management of personal agroforests or multistory gardens has been most heavily influenced by economic and social factors such as crops that can be sold at market or utilized in the household. In many ways, community food forests evolved out of the forest garden tradition, practiced first as home gardens for subsistence and personal enjoyment and now taken to public spaces to provide goods and benefits for the community.

The term *food forest* appears most often in Afro-Caribbean culture. There, annual and perennial mixed farming systems are often used on steep hillsides by peasant farmers or as provision grounds close to the home. At their core, food forests are home gardens with a diverse selection of perennial species that change over time in response to natural succession and community needs. Variations in the availability of light, water, and nutrients for any given species leads to the removal and regeneration of select species over time. Successional patterns, crop yields, and species configurations in a diversified polyculture farm are the agroecological dynamics that agroforestry research spells out and brings to bear on community food forest development.

In temperate climates with limited growing seasons, as the canopy fills in over time and the understory becomes shaded, food forest composition will become less diverse and generally stabilize (though ecosystems are always dynamic and changing). This happens because it is difficult for plants to establish in shaded areas before the growing season ends. The benefit of this is that management requirements theoretically change from intensive tending to something more hands-off and labor requirements are thus reduced, which frees up time to focus more specifically on civic activities.

Charles Kingsley wrote an account of food forestry in the 1870s, following his trip to the Caribbean islands. Kingsley was amazed at the variety of species growing together on provision grounds: mango, breadfruit, avocado, orange, banana, plantain, guava, sapodilla, maize, castor bean, Job's tears, arrowroot, yam, sweet potato, pigeon-pea, granadillo, and many others, including a wide variety for medicinal purposes.[4] Kingsley noted that these food forests provided multiple benefits when used on marginal lands such as hillsides or abandoned plots. Originally, researchers saw these farmers as technologically impoverished, but beginning in the 1960s, research increasingly identified mixed cropping systems as a viable practice. These food forests buffered the risk associated with annual agriculture. They had environmental as well as community benefits including minimizing soil erosion, buffering crops against pests and diseases, and

Forest Gardens

Agroforestry and permaculture definitions of a forest garden are very similar. Agroforestry forest gardens are "reconstructed natural forests, in which wild and cultivated plants coexist, such that the structural characteristics and ecological processes of natural forests are preserved, although the species composition has been adapted to suit human needs."[1] A permaculture forest garden is "an edible ecosystem, a consciously designed community of mutually beneficial plants and animals intended for human food production."[2] Regardless of which definition one is drawn to, forest gardens in either domain "mimic forest ecosystems, those natural perennial polycultures once found throughout the world's humid climates."[3]

increasing economic stability and community interaction by providing continuous sources of food and cash throughout the year. The community food forests of today are likewise designed to produce food that meaningfully benefits the community. Because of this, they are usually fairly elaborate and reflect unique social and site characteristics.

PERMACULTURE IN COMMUNITY FOOD FORESTS

Permaculture was developed in the 1970s in Australia. Since then, it has spread worldwide through associated guilds and institutions. Bill Mollison is widely considered the founder of the movement, although he worked in close collaboration with David Holmgren to develop permaculture's principles. Together, they outlined an adaptive approach to land use that worked with changing environmental conditions to sustain food production. They advocated for landscape design that follows observable patterns in nature. Integrating practices and overlapping functional land use is central to permaculture and practices can occur in something as small as an herb garden or on something as large as a mountainside. It is an intentional, whole system approach to land use that is adaptable to any scale.

Permaculture has more recently grown into a cultural phenomenon. It has evolved into a lifestyle and worldview premised on human cooperation and the ethical care of people and the earth. It addresses consumption, sustaining healthy environments, and developing community. A clear connection between community food forests and permaculture exists. Permaculturists have been central in developing community food forests through public engagement and training because the field depends on knowledge transfer using hands-on learning and theoretical workshops, which are central to most projects. Another way that food forest installations commonly tie into permaculture is through demonstrations of methods that people can replicate at home.

Permaculture practices also have a strong influence on community food forests in large part because edible food forest design is quite advanced within the field. Volumes of detail are available about creating diverse and resilient food forests that replicate species complexity in a young forest through techniques like stacking or seasonally available plantings.[5] The idea of stacking, or layering, in permaculture refers to intercropping species according to height, shade tolerance, and water needs. Stacking can also mean selecting or planning for a plant or element to serve multiple functions. Other key contributions include companion plantings and projects such as raised-bed gardening or free-range fowl tractors that can be moved throughout a site to help control insect populations and provide chicken manure as fertilizer.

Zones

A fundamental aspect of permaculture includes thinking in terms of management zones that radiate concentrically from a focal point such as a home. As one moves away from the focal point in any direction, land use intensity declines in a corresponding fashion and thus overall efficiency increases. A feature that is visited regularly, such as a garden, is intentionally placed closer to the focal point than features that require less oversight, such as a pasture. Another strategy is to purposefully utilize the energy present at a site in the forms of wind, sun, and water to place elements where they will work best with nature rather than struggle against it. These guidelines help with the placement of both living and non-living things across a farm or garden.

The idea of permaculture zones is straightforward and provides useful rules of thumb for where to locate features like a windbreak, small pasture, orchard, house, pond, or beehive. These guidelines help plan for management intensity, water availability, and natural threats (such as a wildfire) and enhance desirable outcomes. In a community food forest, the focal point might be a community gathering space that has plants that require more care and maintenance located around the edges, with trails leading to specific planting areas such as a culinary or medicinal herb garden or a fruit and nut grove. Other examples are areas or structures placed farther away from the gathering space such as a pollinator garden or a reflection bench.

Sectors

Another key design consideration in permaculture is how the pathways intersect with and enhance the composition of desirable features. In designing paths, it is equally important to determine the direction from which threats to productivity and safety come and to learn where preferred things such as wildlife, pleasant views, and sunlight might originate or pass through. Once these items and their direction are determined, features such as windbreaks, fire breaks, or wildlife food plots can be properly located as well. This kind of thoughtful placement of elements is evident in many community food forests.

In a public space like a community food forest, designing to increase safety and security is important for attracting people to the site. The illumination radius of an existing streetlight near the site might be an important sector consideration when designing pathways that will be safe at all hours. Noting where a nearby bus stop is located can indicate edges where hardier plants should be placed that will encounter heavier foot traffic. Another example could be mapping the prevailing wind conditions that tend to blow dust or street litter onto a site so as to plant barrier vegetation instead of edible species in that location. In the Wetherby Food Forest described at the end of this chapter, it was noted after installation that part of a frisbee golf field was located near the pathways, which caused one section of the food forest to frequently experience high impact from players running in search of their frisbees. Luckily, the interference is occasional, but the lesson learned was to map adjacent game fields that might affect the choice of whether to plant in that area or build structures such as a toolshed.

A critical component of sector mapping is observing and mapping water sources and flow through the site and areas of possible retention. Designs use topography and earthwork patterns to capture and distribute water to specific areas or to manipulate drainage. Permaculture promotes reticulation systems of gravity-based catchments and swales that collect water and direct its flow. For example, the Fargo Forest Garden in Portland, Oregon, was established on what was once a parking lot. To recreate the topography of a forest floor, volunteers built up soil, compost, and mulch in mounds around the site to provide microclimates and pockets of moisture.

WHICH APPROACH TO USE?

One way to think about the difference between agroforestry and permaculture is that agroforestry is a discipline and permaculture is a worldview. Agroforestry is a set of scientific practices designed to achieve financial and

PLATE 1. In contrast to rows of crops in traditional gardens, food forests are configured in polyculture patches consisting of multiple species that grow well together. Patches such as this one at the Rahma Free Edible Snack Garden in Syracuse, New York, include both edible and nonedible species. Nonedible species are most often selected for beneficial functions, such as nutrient accumulation or pollinator habitat.

PLATE 2. Fruit or nut trees like this apple tree at the Bloomington Community Orchard in Bloomington, Indiana, are typically planted as canopy species. Shorter plants (shrubby perennials, tall herbs, or flowers) and ground covers are installed below trees or along the periphery of a patch depending on shade tolerance.

PLATE 3. Some community food forests are finding unique ways to incorporate mushroom production into the site design. At the Beacon Food Forest in Seattle, a mushroom hut was constructed and covered with hops to create a shaded area for logs inoculated with shiitake spawn.

PLATE 4. Polyculture plantings are designed to connect across a site until the landscape resembles a young woodland, as shown here at the Beacon Food Forest. This site receives a great deal of precipitation and can support abundant varieties of fruit trees and other species grown in close proximity.

PLATE 5. Community food forests usually consist of native varieties. However, several projects are experimenting with species that are new to the area or previously thought ecologically unsuitable. The food forest at the Throop Unitarian Universalist Church in Pasadena, California, includes plants suited to the dry conditions there, including lavender, sages, nasturtiums, and zonal geraniums.

PLATE 6. Gathering places at community food forests provide space for visitors to comfortably spend time on site and share experiences together. These social spaces do not have to be large or expensive, as the example here from the Mercy Edible Park in Philadelphia, Pennsylvania shows. Brainstorming creative ways to make the space out of locally available resources is an activity in and of itself which can bring volunteers together.

PLATE 7. A newly planted community food forest may appear sparse because trees and woody perennial species are planted quite far apart. Over time as the plants grow, tree crowns spread and ultimately create dense canopy cover, as seen here at the Fargo Forest Garden in Portland, Oregon. Viewed from a nearby rooftop, combinations of mature polyculture patches begin to resemble a forested parklike setting, protecting soil from erosion during heavy rains and providing habitat for birds and other wildlife.

PLATE 8. At the Fargo Forest Garden in Portland, Oregon, mounds of soil were constructed across the site to replicate forest floor terrain, as well as to create microclimates and soil conditions favorable for selected plants. The mounds also help direct water runoff into pathways engineered to channel water to a gravel patio designed with a dry well below to collect water during major storm events. The artistic gateway improves site aesthetics and attracts visitors.

PLATE 9. At the Freedom Square Food Forest in Troy, New York, the Sanctuary for Independent Media hosted a community workshop to create a mosaic mural stage with a world-renowned mural artist, Isaiah Zagar. More than 100 community members attended the event and contributed photographs of those lost due to violence and sayings that reflect community values on peace, justice, and growing their own food. The stage is used and respected by the community for all kinds of events, from music concerts to peace rallies.

PLATE 10. A large bee sculpture at the Beacon Food Forest in Seattle calls attention to the importance of habitat and forage for pollinator species. The composting bins nearby are creatively designed in the shape of a honeycomb. These artistic site elements provide talking points that can help people learn about how food forests support pollinators.

PLATE 11. The Fargo Forest Garden in Portland, Oregon, is located next to a café. In addition to providing culinary herbs for use in specialty drinks, the community food forest setting offers patrons a relaxing dining experience. A wooden sign and metal-sculpted insects on the gate reinforce the role of community food forests as environmental habitat.

PLATE 12. A fence around a food forest can appear to be a barrier, but at the Basalt Food Park, in Basalt, Colorado, the fence provides a way to post bilingual information informing people about the project and inviting them to participate. Translating signs can be done with the help of bilingual volunteers or by partnering with multilingual community groups. At the Hale-Y Community Food Forest in Blacksburg, Virginia, signs were translated into nine languages by partnering with a local YMCA international women's group.

PLATE 13. Signs at community food forests are critical. They range from detailed descriptions about topics such as ecosystems and food production to something simple like those that identify plant species. When funding allows, it is best to invest in weather-resistant and durable signs like the example here from Beacon Food Forest. This sign is made of tile, and the information is translated into multiple languages.

PLATE 14. Signs that include maps help visitors find their way around, know where to harvest, and identify who is responsible for the site. This sign at the Beacon Food Forest is made with chalkboard paint so that harvest information can be changed when needed. It is also a way to indicate areas that might need maintenance.

PLATE 15. If funding is not available to create permanent signs, semipermanent signs can be quickly prepared at low cost using poster board or wood with oil paint sharpies or outdoor acrylic paint. Locating signs next to a gate helps publicize that the site is open to everyone. When the budget allows, longer-lasting signs can be placed throughout the food forest.

PLATE 16. Small signs or labels on individual plants that include information on when and how old a plant was when transplanted is helpful for gauging when to expect the first harvest.

PLATE 17. Here is an example of a wooden sign painted with outdoor protective paint and oil sharpies at the Hale-Y Community Garden Food Forest. Volunteers made the signs during an educational workshop where participants selected a plant to learn about and added researched information to the signs, which are located along a walking path next to the appropriate species.

PLATE 18. Some community food forest sites invest in kiosks or covered boards to create a place for posting information about the site, events, community meetings, and more. Boards can also serve as a vehicle for community communication where notifications that are not site related but relevant to building community are posted.

PLATE 19. The center of the Bloomington Community Orchard in Bloomington, Indiana is covered by a large patch of vegetation where visitors can walk among tall feathery shoots of asparagus and strawberry bushes. This site offers a variety of experience from open areas in full sun to more shaded areas where there are mushroom production experiments. Including a variety of experiential conditions in the design invites exploration and curiosity.

PLATE 20. In Asheville, North Carolina, most of the canopy species of the Dr. George Washington Carver Edible Park, such as walnut, pecan, apple, and jujube, have reached mature height and fullness, creating a shaded understory in the summertime. The shade provides a comfortable, cool place to gather and rest, but a balance must be kept to avoid cramped or dark spaces, which some may view as unsafe.

PLATE 21. The 18th and Rhode Island Permaculture Garden in San Francisco, California is located on a 14 percent slope. Rock supported terraces and mulch pathways built on the contours help manage water and store it on site. The terraces also raise food growing areas away from the path and small signs like this one provide a reminder to clean up after pets. Encouraging people to wash produce before consuming it is also a best practice.

PLATE 22. At the Basalt Food Park in Basalt, Colorado, students on a field trip learn about the food forest.

PLATE 23. Including an area in a food forest where children can play gives parents the opportunity to talk informally, create social bonds, and potentially discuss involvement with the food forest.

PLATE 24. Strategically placing buildings and other infrastructure can be advantageous, as demonstrated at the Bloomington Community Orchard in Bloomington, Indiana, where a shed was designed to collect rainwater off the roof while also blocking wind to provide a warm gathering space for volunteers harvesting apples in the cool fall months. The rock slabs enhance the gathering space and radiate warmth for the small fig tree planted at the center to help it endure cold winters.

PLATE 25. Gathering spaces do not have to be elaborate to be effective. Something simple like this rustic example from the Basalt Food Park in Basalt, Colorado, invites people to pause and spend time in the space, increasing the chance to observe and connect with nature.

PLATE 26. Gathering spaces are one of the most important elements to include in a community food forest. They range from a simple sitting area like that shown in Plate 25 to elaborate setups for workshops and food processing, like this central gathering hub at the Beacon Food Forest in Seattle, Washington.

PLATE 27. In places like Santa Barbara, California, it is essential to select shaded areas or create sheltered spaces where volunteers can cool off after long days of work. Each gathering space is unique, evolving over time with additions from volunteers or community groups.

environmental goals. Permaculture practices are largely based on environmental observations and ideas about nurturing relationships between the land and people. Using one or the other approach when designing a community food forest will lead to slightly different planning and management decisions.

So, what do agroforestry and permaculture look like in a community food forest? Permaculture can provide vast and growing stores of human energy in the form of volunteers and project leaders from broad networks. Many people practice permaculture, and it continues to grow in popularity. Use of permaculture practices such as forest gardens, water catchment (a piped water system), and zones and sectors can shape the overall layout and visual quality of a community food forest.

Agroforestry provides the science that can help with technical design and maintenance strategies of a food forest project—how to manage combinations of trees, shrubs, and crops. Some agroforestry researchers are interested in studying community food forests, and their involvement can help increase the competitiveness of a project when it comes to winning technical grants and assistance. Much of the information taught through permaculture comes from scientific fields, but only in the last few years has formal research been initiated to better understand the practices. Agroforestry, on the other hand, has a robust record of researching plant interactions in order to suggest detailed management plans and economic viability.

Permaculture design places emphasis on the relationships between plants as well as their resources, such as soil-to-plant interaction. Permaculture design principles are helpful in organizing site layout in an effective, efficient, and accessible style. Existing site elements and ease of use help determine placement of recreation areas, spots for community activities, and food-production areas. Learning about agroforestry can help project managers better understand ways to manage plant and tree interactions.

Combining permaculture and agroforestry in the planning, design, and management phases of a community food forest allows project leaders to create a project vision that is both practical and competitive from the standpoint of human resources, project funding, long-term maintenance, and policy making. The extent to which the two can be assimilated depends upon the nature, needs, and backing of a community food forest project, but project leaders should consider the role both can play in achieving their goals.

Both disciplines are adaptable, and adaptability is critical in a community food forest because the demands of intercropped trees and plants change over time across unique site conditions. Permaculture design guidelines like zones and sectors help in deciding where to place particular plantings, built elements, or other features. After locating the best places for specific sections of a design, agroforestry guidelines can be used to determine specific botanical configurations that have been tested and are known to produce results. For example, deciding on the location for a barrier to wind, dust, or street debris uses permaculture guidelines. Agroforestry can provide the specific planting configuration of creating an A-shaped barrier with two rows of short, thick bushes divided by a row of taller trees between them. Accompanying calculations estimate the area that will be protected from wind based on the height of the tree row. Additionally, agroforestry provides details on potential maintenance needs and management plans, such as pruning and thinning.

The differences between the two disciplines can also be exploited to support the practical facets of a community food forest. Knowledge of agroforestry can help when defining the purpose and composition for particular plantings. It can fine-tune relationships between species with science-based management. Permaculture can aid in conceptualizing the design for particular plantings using social factors to help determine species. For example, one area might be planted with species that are used by a specific ethnic population to create a space where they feel invited to harvest and celebrate.

Permaculture is also rich with design formulae and palettes drawn in compelling detail, with clear influence from the landscape architecture field. Locating a community food forest on public property sometimes requires approval of the design by a certified landscape architect to meet city regulations. A detailed permaculture drawing can make it easier to share site design concepts with landscape architects because it uses a format familiar to them when finalizing plans to meet city codes and the community's vision.

Though there are differences between agroforestry and permaculture, clear similarities they share are a fundamental emphasis on the whole environment and incorporation of traditional ecological and contemporary ethnobotanical knowledge. Both also integrate trees, crops, and livestock on the same piece of land and both recognize potential connections to people, places, and community development. Both promote beneficial plant communities that minimize competition and distribute resource demands, recycle nutrients, fertilize the soil, and effectively grow together over time. They hold the attention of progressive people seeking to creatively design sustainable and resilient land use systems at various scales. The Wetherby Edible Forest in Iowa City is a useful example of an adaptive approach that integrates agroforestry and permaculture.

The Need for Detail— and for Flexibility

Anecdotal evidence indicates that favoring detailed technical design in the planning phase of a community food forest and rigidly adhering to that design during establishment will likely threaten project success. Many community food forests leaders learn that flexibility is necessary if a project is to succeed and become meaningful to the community. Permaculture design relies heavily on selecting and integrating plants that fill ecological niches designed to imitate a forest ecosystem. Investing time and energy into assessing the site, measuring the spacing of plantings, and creating arrangements of species that work together to re-create a forested landscape is a worthwhile exercise. It is essentially due diligence toward thoroughly understanding the potential and limitations of a site and project design. However, regardless of how much time is put into the initial design, it is important to remember that changes will inevitably occur as plants and community involvement grows and evolves.

THE WETHERBY EDIBLE FOREST

The Wetherby Edible Forest in Iowa City was created in 2011 through a grant partnership between a nonprofit organization called

FIGURE 8.2. Although the Wetherby Edible Forest is only four years old in this scene from 2015 (periphery plants are only one year old), the structural patterns of a small forest are evident and beginning to blend with the surrounding area. Note the lined pathways that give a visually aesthetic cue that this area is purposefully designed rather than a random patch of perennial species.

Backyard Abundance and the Iowa City parks and recreation department. The food forest began as an orchard-inspired edible maze on barely a sixth of an acre. The goal was to increase educational opportunities at parks in the city and introduce youth to local urban food production.

Fred Meyer formed Backyard Abundance to help people design and install environmentally friendly and aesthetically pleasing projects that provide food and enhance biodiversity. The organization holds public seminars, hands-on training, design consultation, and public demonstrations throughout Iowa City and surrounding areas. By 2014, the Wetherby Food Forest had doubled in size and transitioned into a full food forest focused largely on agroforestry education.

Fueling the growth of the Wetherby Food Forest was the financial support received from the Iowa Department of Agriculture and Land Stewardship Specialty Crop Block Grant Program. Volunteers installed edible perennials as a demonstration for regional farmers as well as a base for pollinator habitat. A key deliverable of the specialty crop project was the *Edible Agroforestry Design Templates Manual*. It showcased replicable designs and provided instructions for site assessment. It also made species suggestions for alley crop orchards, edible forest edge projects, shady edible forests, edible riparian buffers, edible windbreaks, homestead orchards, and silvopasture. The manual is free and available to download at www.backyardabundance.org.

Permaculture is a source of inspiration at Backyard Abundance and a driving force behind initiatives such as the Wetherby Edible Forest. Agroforestry found its place in terms of partnerships and publications for specific land use demonstrations and technology transfer at the site. Farmers learned about agroforestry, which helped create a clear connection to institutional funding sources through its science-based practices. Altogether, this example not only showcases the possibility of generally blending permaculture and agroforestry but highlights a specific way these practices can be simultaneously leveraged in support of a community food forest and associated organizational partnerships.

The history of the Wetherby Edible Forest shows that the question of scale is a key consideration for sustainability. In hindsight, Backyard Abundance and Iowa City Parks and Recreation, which recruit volunteer maintenance for the food forest, determined that one of the most important lessons learned was to start small and grow over time. In order to receive funding for agroforestry demonstrations, the site had to be at least one third of an acre, but following through on developing even one sixth of an acre proved difficult in light of the extent of community support they were able to generate. With an honest assessment of capacity in mind, an organization or collective of community members working on a community food forest project can selectively pursue grants with realistic and achievable requirements.

CHAPTER 9

Allies in Creating and Managing Public Space

People gather, exchange ideas, and trade goods and services in public space. It is where cultural, political, and economic expressions are possible. The role of community food forests in shaping public space must be thoughtfully considered. Since community food forests are sites for public demonstration and education, they can be examples of values, perspectives, cultural beliefs, politics, economics, social structures, food production, and civic engagement. Leaders must not take such a responsibility lightly.

Like political discourse, the educational opportunities that are amplified, downplayed, or generally available at a community food forest will reflect the stakeholders who are involved. A constructive environment is critical. Thus, the first part of this chapter is dedicated to understanding public space and how it can be a medium for expression, as well as the ways in which the landscape serves as a form of language that provides information about the places we share with others. In that regard, an array of professions and people engaged in the business of designing and maintaining public spaces can be allies in the design of a community food forest.

In the second part of the chapter, we consider how potential public space allies and other public community members play a role, whether active or passive, in the design of community food forest projects.

As we visited sites and talked with food forest leaders, it became clear that many did not know where to turn for information on how to realize their vision for a community food forest that would serve as a useful and enjoyable public place. They either did not know where to find the right information, or—more important—they were not sure who their allies were or which professionals could help them and how. For example, many leaders did not know how to care for fruit trees or who could teach them the necessary skills. Others had no idea how urban foresters could help with a community food forest or even that such a profession existed. Some leaders thought local government professionals might be against the idea of a community food forest outright. Partnerships that help package ideas to improve chances for acceptance and success in the public sector are much needed.

There are many professionals who can contribute to a multifaceted community food forest design that safely and sustainably complements public space in urban environments. Any one professional group on their own should not be counted as a singular solution when designing a community food forest project, but all should be considered and involved. Because the publications and resources pertaining to these professions are extensive, in this chapter we go only so far as to introduce a few of the most relevant and point out some aspects and potential services worth considering when designing a community food forest. Management and design studies on urban parks, gardens, and forests can provide useful ideas for scaling up a food forest from backyards to functional and safe community places. Intersections between the professions we highlight may happen naturally or may need to be the product of a targeted effort to figure out who to enlist to achieve defined goals, and in what capacity.

Integrating the design, installation, and maintenance of a food forest into the day-to-day life of a community is easier said than done. Building a project that is conducive to interaction across age, identity, and culture over time requires careful thought about how to serve the community through relevant planting strategies, useful infrastructure, and community programs. It is also important to understand that a young forest eventually grows into a shady grove or a dense and sometimes messy wooded area. Changing conditions in vegetation must be managed for usability but also for concerns such as public safety. Urban agriculturalists and gardeners can help enhance short-term yields at community food forests by integrating annual crops. Professionals who study forests and understand natural succession and the needs of managing maturing trees can be helpful in creating long-term plans.

UNDERSTANDING PUBLIC SPACE

The term *community food forest* is often thought to signify a location on public property, but even when community food forests are located on private land, they are almost always open to everyone. Ownership is not the only defining characteristic of public space. Of course, public space can be a physical location where people meet, but it can also be a newspaper where society communicates. It can be even less tangible, such as people engaging in expressions of politics, social change, and ideas that affect civil society, no matter where and how such discourse happens.

> ## Community Food Forest Interaction
>
> As community food forests grow in number, so too will their reach and relevance. This will increase the need for community food forest leaders to look beyond their local projects and network with others through regional meetings and collaboration, and perhaps even through national conferences, where ideas and energy can be exchanged and leveraged. Such gatherings build awareness and inspire thought about community food forests. Topics addressed may include the types of changes in conditions, best planning strategies for managing yields, professions to partner with during particular project phases, and examples of how community is acted out and shaped in these public spaces. Information is shared for the benefit of all involved in their creation and care. Additionally, leaders and participants from various sites can help each other learn about how to handle internal conflicts, thereby supporting volunteers and other food forest community issues that we discuss in Part 4 on building meaningful community.

Public spaces reflect the values, identities, and action of those involved within a community. A statue, a garden, or even the design of a building can be a communication tool in the public realm. A statue might provide clues to the history of a place; a well-tended garden can communicate that a space is cared for on a regular basis. Understanding how public space is used, thinking about who will feel invited into it and why, and being deliberate about the types of community resources provided are all important things to think about when planning a community food forest. Anticipating how the broader community will perceive involvement in your food forest project can influence who participates as a volunteer. Michael Diamond, a psychoanalyst who writes about humans in public space, notes that the way we experience, use, and define public space is socially, culturally, and individually influenced.[1] It is space where we encounter "the other." Many community food forests aim to demonstrate that citizens are stepping up to actively design and test civic solutions rather than waiting for others to provide answers. Actions and words used in public space become part of the public dialogue about food production, urban agriculture, urban land planning, and other controversial topics. In fact, it was brought to our attention by one project leader that part of their motivation to create a community food forest was to join the national conversation about food security and community.

Landscape as Language

Community food forests can be found in a public park located on parks and recreation property, a neighborhood common, a streamside planting, or a formerly vacant lot. They are edible landscapes that border recreational trails, a reflection garden on church property, or an outdoor classroom at a school. In every instance, they are public spaces and forms of communication. However, community food forests are not conventional forms of manicured public spaces or even typical community gardens that most people are accustomed to. Some people may reject unconventional landscapes at first if they do not understand them. Perceptions of acceptable landscapes are formed by cultural and socioeconomic backgrounds and transformed or refined over time through exposure to new sights, ideas, or education. This reality shapes people's experience, engagement with, and acceptance of community food forests and urban landscapes in general.

Our cultural aesthetic preferences are sometimes rooted in policies that have become an embedded way of life without question. An observer of a community food forest may not understand that they are looking at an ecologically functional food production system, or they may not accept it because it is unconventional.[2] Many people expect that even "natural" landscapes in inhabited areas should look well maintained; "wild" and "messy" landscapes should only exist in uninhabited areas. The result can be a cultural barrier that deters some from entering a food forest based on its appearance or a lack of framing to place the food forest into a context they feel comfortable with and understand.

This conundrum is explored and explained by professor of landscape architecture Joan Iverson Nassauer, who puts forth the notion that the goal of using aesthetics in an ecological landscape design should be a strategy of using the local language (accepted aesthetics) for introducing new concepts (perennial agriculture with multiple vegetation layers) in a way that people understand (such as confining it within mulched pathways).[3] Aldo Leopold understood this same concept when he observed, "Our ability to perceive quality in nature begins, as in art, with the pretty."[4] In other words, when designing public places, it is important to take into consideration cues that signify a landscape is looked after and orderly. Patches of food forests can be lined by clear pathways or low walls, entrances and exits should be clearly marked, and general aesthetics must be thoroughly considered.

The Community Food Forests Context

In a traditional sense, food forest design focuses on botanical structure. This normally includes planning for the vertical arrangement of select plants and trees into multiple layers, along with how they should be distributed across a piece of land. Designers should also consider how plants will interact with each other over time. In some cases, the site design will need certification by a landscape architect; in others, it may need the approval of the town horticulturist or urban forester. The physical aspects of design are important, but community food forests are human spaces in a particular place surrounded by unique community culture. It usually "takes a village" to help blend the social and biophysical aspects to create spaces that will attract visitors and facilitate interaction with the site.

The composition of the stakeholder group will be unique for each community food forest. However, when visiting sites across the country, we noticed that people from several key professions were often involved: landscape architects, urban designers and planners, urban foresters, urban agriculturists, and urban gardeners. In addition to these professions, there were frequently people with various titles from government agencies such as parks and recreation departments, sustainability offices, public works, and department of neighborhoods offices. We recommend thinking strategically about how to involve these professions and agencies as partners in your community food forest.

LANDSCAPE ARCHITECTURE

Many community food forests have needed certified landscape architects to sign off on plans or found benefit in having a landscape architect involved in the design process and site development. Landscape architecture is a multi-disciplinary field that integrates elements from the natural sciences, sustainable management, industrial design, community development, environmental engineering, and urban planning. Landscape architects harness the knowledge of ecological systems to design plant communities in accordance with specific goals or standards. In this sense, there is significant overlap with permaculture, most specifically with respect to mapping and planning.

The basic permaculture design method is made up of an initial phase for collecting useful information and a second phase when this information is incorporated into a final design used for implementation. The first step in collecting information is visioning and articulating project or design goals. Next is site observation, which entails taking specific notes about site opportunities and constraints that can be examined during the following step. After observation there is a thorough site assessment, which includes creating a base map of site conditions, characteristics, natural and manmade features, and mapping out important living and nonliving elements affecting and shaping the site.

Some people combine assessment with site analysis, but assessment can also be thought of

ALLIES IN CREATING AND MANAGING PUBLIC SPACE

FIGURE 9.1. A landscape architect may use asset mapping with a community and then apply the knowledge to decide where best to locate a community food forest and pathways connecting important places in the neighborhood. Here, a final rendition shows a community garden area integrated with a food forest and perennial riparian plantings in the lower right corner. Illustration by Will Serge.

as a separate step that analyzes all the information collected through goals and observations. After the analysis, an initial concept map is roughly sketched along with others that show potential alternative configurations. Feedback is then collected from all involved. In the case of a community food forest, feedback is generally gathered from as many stakeholders as possible, especially the public. It is then incorporated into the design before a second detailed design is created. This more specific design uses exact measurements for all elements and comprehensive species information.

A final design is often referred to as a plan view. Landscape architects use a survey or an aerial image of a proposed site as the base and then utilize software and botanical knowledge to layer proposed plants, hardscapes, and other infrastructure over the map at a scale correlated to site parameters. A good design also includes implementation details, such as where to source materials, estimated budget with cost analysis, and instructions for immediate, short-term, and long-term care.

When it comes to community food forests, landscape architecture design should always take into consideration public opinion for increased sustainability of the site. Public design charrettes are a great technique for explaining the concept, collecting opinions, and responding to feedback.

This cycle can happen numerous times until an agreed-upon design is finalized or as budgets allow. Due to the scale of community projects, a design team is usually assembled that can inform project design with site and community surveys, as well as assist with policy and code compliance, such as making sure a community food forest's paths are compliant with the Americans with Disabilities Act.

Landscape architects give equal attention to the infrastructure needed to facilitate use of the site, such as lighting to allow the area to be safely visited at night. Hardscaping needs such as grading for walkways, platforms for sheds or pavilions, and material selection for pathways and patios can be estimated along with calculating approximate loads of surface water runoff. Rather than producing one detailed site design at the end of the process, there will be a master plan that includes a series of designs that visually represent the relative location of each element on the site as well as different visual perspectives of the project. (See figure 9.1 versus 9.2.) Tree crowns and other vegetation are represented at mature size. Pathways are shown, and their material, style, and color may also be symbolized.

Growing Goodwill Community Garden

Another layer of detail that landscape architects bring to the planning process is determining how a community food forest can connect community

FIGURE 9.2. Landscape architects present a series of designs for a comprehensive explanation of a site plan. A close-up view (*left*) provides details regarding specific site layout. (See figure 9.1 for the location of the food forest in relation to the neighborhood.) Illustration by Will Serge.

ALLIES IN CREATING AND MANAGING PUBLIC SPACE

at the landscape scale. They are equipped to help communities visualize the design space through illustrations and drawings. In Roanoke, Virginia, Goodwill Industries of the Valley and the Roanoke Community Garden Association worked together to create a community garden on part of Goodwill Industries' property, which was a vacant field. The location of the garden made fresh food accessible to Roanoke's Shenandoah West neighborhood, which includes the city's largest public housing complex. Half of the residents of Shenandoah West live below the poverty line, and food security is an important issue.

When the garden association sought to do more with the site, their director, Mark Powell, approached Will Serge, a landscape architecture student at Virginia Tech, about taking on additional site design in support of the association's aspirations. Will adopted the project for his thesis. In addition to assessing the site itself, he studied surrounding neighborhood features, surveyed the residents of the public housing complex, and noted other important elements in the area that could be integrated into site design. Results showed the garden was located on the floodplain of the Goodwill property downslope from a small grove of trees and in proximity to an underutilized park and the local library. Since many of the residents did not own a car, walking was the main means of transportation between points of interest around the community garden and complex area.

Additionally, landscape architects use visual simulation skills to create dimensional perspectives that, as seen on the right, can be presented to a community to convey a vision of what the food forest will look like after the plantings have matured. Illustration by Will Serge.

Community partner organizations in Roanoke had previously worked with residents of the Shenandoah West neighborhood to develop a transformation plan as part of the US Housing and Urban Development Choice Neighborhoods Grant program. The plan provided a comprehensive design framework for Will to build on using the information he collected to meet the goals of the grant as well as to fulfill the missions of both Goodwill Industries and the Roanoke Community Garden Association. Will's approach to planning and design was inspired by Christopher Alexander's *The Oregon Experiment*.[5] He fashioned a participatory process based on methods described in the book and held three community meetings over the course of a year to introduce the project, gather data, present the conceptual design, receive feedback, and host a final meeting to share results. A design team of eight participants representing stakeholder organizations and residents met every two weeks between the community meetings to refine project vision and details.

This comprehensive process created a steady dialogue that enriched Will's understanding of the community and stakeholders, as well as of the areas in the neighborhood that were valued, constrained, or offered opportunity for design. Along with other features, a food forest emerged as a project that could help add trees to the sloped areas of the property above the garden while providing an additional source of food, a recreational area, and an aesthetically pleasing greenway that connected the library, parks, and community garden for a more enjoyable experience. This level of planning and participatory process gave residents a voice in the final design and created a space where their values were reflected from the beginning. It also embedded the food forest within a larger contextual plan by linking it to surrounding features and elevating its role in the community.

URBAN DESIGNERS

Urban designers are usually concerned with the purpose and configuration of public spaces for people, animals, and plants, such as public parks and plazas, botanical gardens, urban forests, and community gardens. Designers think about how relationships between people and their surroundings impact local culture. Designers aim to reflect a community's values by creating spaces that reinforce but sometimes also challenge them. Urban designers help connect people in built settings, working to ensure alignment between key interests and the structure and function of on-the-ground space. Design professionals can help create a food forest plan that will attract people and provide a valuable experience for them. They can help answer questions such as: What will help draw people to food forests? Will they enjoy the visual and emotional experience once there?

Some community food forests use vegetation in artistic and appealing ways to improve site aesthetics. They often do this either through pathway design that ensures flow through a site, installation of circular or organic designs, or plantings of ornamental and colorful varieties along walkways or entrances. Others employ artistic entranceways, sculptures, or colorful signs to signify the food forest is also an art garden, which increases its value as a public asset. (See color plates 10 and 11.) Another focus is the role of architecture and organization in social use—how individuals interact with and perceive public space. This element is centered within a new perspective focused on place making, which involves designing visually appealing places and aesthetically unique projects, such as green amphitheaters, that also serve social and environmental functions.

City landscapes have evolved over centuries as urban designers keep elements that work, reconfigure those that do not, and experiment with new ideas intended to improve the overall well-being of urban residents. In the 1960s, William H. Whyte studied how people design urban spaces, and many of his insights can be used today to guide food forest design. His insights include the prudence of designing and maintaining something highly attractive and inviting so that people are drawn to use the site. An increased number of visitors translates to more people monitoring activity, increasing a sense of security. This is a way to minimize undesirable activity through social means rather than something obstructive like fencing: "The way people use a place mirrors expectation."[6]

Whyte's observations include insights on eating arrangements and styles; site capacity; the ways in which people gravitate to specific areas depending on natural elements such as sun exposure, trees, wind, or water; and the kinds of stimuli that prompt people to interact with each other and increase the sense of safety in a location. When designing a food forest that will encourage and discourage use when and where preferred, community food forest project leaders would benefit by looking closely at Whyte's observations. Project for Public Spaces, a nonprofit organization based on Whyte's work, focuses on planning, design, and education dedicated to transforming and creating public spaces that build stronger communities. The organization is a central hub for the placemaking movement in urban planning and design, and their website offers multiple resources that can help build a sense of place through community food forest design.

Creating a clear understanding, whether blatant or implied, about where people are and are not allowed to go in public space can obviously profoundly impact rates of use. The extent to which individuals can and, more important, feel able to leave their imprint on a place should

Sense of Place

Experiences with particular places are rarely strictly physical. The characteristics of a place can evoke certain feelings—a mood, nostalgia, or what is referred to as a "sense of place." Sense of place emerges from how people subjectively construct meaning out of their experiences, and urban designers are very attuned to this phenomenon. Designers blend landscapes, activities, social interactions, and other phenomenological elements to transform a nondescript space into a unique, emotive place.

Community food forests seek to provide a place that many will embrace, whether they want to be actively involved or passively enjoy what the site offers. Including community members in the design of community food forests increases their sense of connectedness, which in turns means they will be more likely to use and steward the resulting place. Involving a design professional who is skilled at seeking input gives community members a voice and can strengthen their desire to implement and maintain their food forest.

also be considered when creating a community food forest. Designating which areas are open for planting by those who want to add plants from their home landscape is important if a site's goals include encouraging public participation. Another way to encourage participation is to designate areas where people can make a mark through creative gathering and sharing. Regardless, thinking about where and how to designate access is important, as is considering how certain symbols or perceptions may create invisible barriers.

Pasadena, California

In Pasadena, California, Transition Pasadena used the food forest concept to design an edible landscape around the Arroyo Food Co-Op and the Throop Unitarian Universalist Church. The plantings incorporated multiple elements of urban design to enhance attractiveness. The food forest was installed at an intersection that receives a lot of foot traffic and the area around the store was dull and void of any noteworthy vegetation that would encourage community use. Project volunteers sculpted the ground to create water-catchment swales, which also added visual complexity to the landscape. They then planted a variety of edible species (cabbage, kale, strawberries, beans, tomatoes, broccoli, bok choy, and more), three donated fruit trees (tangerine, apricot, and persimmon), culinary herbs (dill, oregano, and others), and native drought-resistant plants. They chose plants that did not inhibit views of the street or sidewalk but added texture and color against the plain backdrop of the building.

The walkway through the planting was designed to slow water runoff, but that design also slowed the flow of people, so a snaking pathway of stepping-stones was added. Added seating invited use of the space, as did signs that promoted picking the edibles and encouraged people to leave their mark in the form of small creative contributions or items they no longer needed for others to take. The group began leaving boxes of surplus fruit on the tables, and then noticed that others began to bring fruit from their backyard trees to share as well. The Arroyo Food Co-Op unfortunately closed in 2016 when the lease expired, but its legacy remains as does the community food forest. The success is a testament to how the food forest concept can be used even in very small spaces to enliven an otherwise drab urban place and enhance public space through infrastructure such as benches and tables that invite people to gather and glean from the site.

Transition Pasadena also transformed the grounds at the Throop Unitarian Universalist Church from a landscape that required heavy irrigation to maintain its turfgrass, roses, and other water-loving bushes into a dryland-inspired food forest called the Throop Learning Garden. Throop Church had received a citation from the City of Pasadena because its weed-filled lot was deemed a public nuisance. The church was also concerned about the rising costs of water and the need to conserve during a drought. At this time, Transition Pasadena was holding its meetings at the church, thus becoming aware of the water challenges, and decided to do something to help. In 2010, they rallied their community to join with church members to remove Bermuda grass and hardpan clay to make space for soil rehabilitation using compost and mulch. As a result of meetings with the congregation and adoption of permaculture and urban design principles, the group established a holistic, drought-tolerant project. Features included raised beds for elderly church members; fruit trees located close to the corner

FIGURE 9.3. A patio made from urbanite (recycled broken concrete) designed by Transition Pasadena members collects water with a French drain design for redistribution to plants at its edges. Local high school environmental science students contributed labor to install the patio. It is made in the form of a spiral where one can stroll in contemplation toward a heart-shaped stone in the middle.

bus stop, so those waiting could grab a healthy snack; culinary and aromatic herbs; drought-tolerant varieties of roses (an important element to congregation members); and native plants, many of which also had medicinal or sacred value, such as white sage, yerba santa, milkweed, and mugwort. By 2013, instead of receiving nuisance citations from the city, their hard work was recognized with a Pasadena Green City Award for Urban Nature.

In the spirit of urban design, there are clearly marked and maintained pathways that allow a visitor to move through Throop's community food forest as they choose. Places to sit and reflect were strategically selected so the garden could also be used as a place of meditation, prayer, and reflection. Site leaders hold earth meditations in the garden and provide educational workshops. The church raised $5,000 to install a rainwater harvesting system on its roof and it has become a learning center for water harvesting techniques.

In 2016, Transition Pasadena installed a hügelkultur bed and a Three Sisters–style (corn, beans, and squash) vegetable garden. The corn grew seven feet tall, but wildlife found it before the harvest. A raccoon family found living in a storm drain across the street enjoyed some of the bounty, proving that even in unsuspecting urban places, wildlife can find their way to the pickings.

FUNCTIONAL DESIGN

URBAN FORESTERS

Urban forestry is a relatively new field, but one that is shaped by centuries of rural and community forestry where local residents help direct management activities. It generally involves the management of wooded areas and plantings of trees at various scales in built environments. Traditionally, urban forestry was limited to tending street and park trees, planting new trees, and collecting and processing urban wood waste. Today, however, the field is multidimensional and interdisciplinary. It still includes conventional tending and maintenance practices, but also tackles landscape issues such as the systematic expansion of urban forest canopy cover and complex social and political processes that affect the health, productivity, and usefulness of urban trees and forests. Many of the concerns of urban foresters are those of community food forest leaders as well, particularly in terms of melding tree-by-tree management with the broader responsibility of sustaining healthy, productive, useful, and safe woodland canopies in cities and towns. Urban foresters and urban forestry could pay vital dividends in the long-term management of community food forests and their contribution to urban forest canopy, water and air quality, and soil health.[7]

The knowledge urban forestry experts have about managing wooded areas and trees in cities and towns can help you develop an inclusive, useful, and safe project that maintains healthy and productive wooded settings. When done well, community food forests become a functional component of the larger urban forest where wild food, medicinal plants, and timber and nontimber materials are not just managed for but are also shared and available without judgment. All of this points to the growing role

Preparing for Conversation with Professionals

Each professional discipline has its own jargon. Each also has its own concerns and ways of thinking about projects. When preparing for a meeting to discuss your project with a professional, there are common references that can help you anticipate the questions or concerns that may arise. *Time-Saver Standards for Landscape Architecture* is a valuable reference and guide regarding those things that landscape architects are apt to consider and how they might discuss trade activities such as planning for natural hazards, stormwater management, pedestrian circulation, retaining walls, sound control, and masonry.

The International Society of Arboriculture website has multiple publications and podcasts covering topics of interest to arborists. The website for the US Forest Service's Urban and Community Forestry Program publishes resources for grants and publications, such as the Ten-Year Urban Forestry Action Plan (2016–2026), which can help you better understand how community food forests fit into the agency's local and national objectives.

Researching websites, publications, societies, and organizations associated with the professions mentioned in this chapter can help improve your chances for making inroads with them. Such an outcome can improve your community food forest planning and identify language and concepts that are appropriate for grants in particular focus areas.

of perennial food and other woodland products in the spectrum of benefits derived by communities from their urban forests.[8] Darrin Nordahl[9] and others make a strong case that a surprisingly large number of people need to forage or glean from urban forests to feed themselves and their families.[10] He additionally describes how policies that allow foraging or gleaning, or even encourage it, can significantly help reduce maintenance and liability issues often caused by antiurban agriculture policies, particularly regarding fruit trees, which are considered messy and vectors for insects and sometimes disease.

Nordahl's descriptions of how urban foraging can help build community and connect people to landscapes are compelling. However, achieving an outcome with potentially profound implications requires first that the biological basis be something stable and useful. Whether messy or orderly in appearance, a community food forest requires maintenance and oversight. Otherwise, project aims and objectives may eventually be lost over time. In this regard, urban foresters can be important professional partners to include when designing and managing a community food forest project because their work improves the physical and psychological health of residents and can substantially enhance community well-being.[11] Recall the nested systems concept discussed earlier in the book. Urban foresters can help you understand how the site will fit into the larger urban canopy system. They should also have information on growing conditions and limitations for the region that could be useful when selecting species.

URBAN AGRICULTURISTS

Urban agriculture gravitates back to localized forms of community food production on public land in the eighteenth and nineteenth centuries. Though this practice was noble in some ways, we should largely be thankful that the work of urban agriculturists today is not a strict replication of 250-year-old commons. The modern form of urban agriculture recognizes the value of its public place in addressing urban and peri-urban community revitalization and well-being. Thus, urban agriculture is a multidimensional practice that strives to supply high-quality food and also to provide community benefits and services such as "recreation and leisure activities, economic vitality, business entrepreneurship, individual health and well-being, community health and well-being, landscape beautification, and environmental restoration and remediation."[12]

After fifty or more years of out-migration from urban centers, a new era is underway with revitalization and in-migration as its hallmarks. All of this is underscored by exploding world population growth and the tipping of scales in 2008, when for the first time more humans lived in urban rather than rural areas.[13] The increase of populations in cities, especially in impoverished and underserved clusters, has vaulted issues of urban agriculture to a level in need of immediate global attention. In response to these changing social and economic conditions, the number of urban agriculturalists is growing, along with innovative technologies and production methods.

Urban agriculturists help to integrate food production spaces into the daily backdrop of the citizens they intend to benefit, such as by building lobbies, outdoor seating areas at restaurants and cafés, or public promenades. Since the early 2000s, more than 1,000 golf courses have permanently shut their doors. Some of these expanses of land have been transformed into agricultural parks and conservation ecosystems.[14] In September 2015, urban agriculturists

FIGURE 9.4. At Mercy Edible Park in Philadelphia, annuals like the "Three Sisters" (corn, beans, and squash) are mixed together with perennials to serve as a demonstration of the ways in which sustainable agriculture can integrate short-term yields with slow-growing perennial plants. In another section of the food forest, traditional rows of lettuce and other greens are placed in front of a fence covered with hop. This arrangement mixes easily recognizable agricultural crop plants with perennials, quickly indicating to passersby that the site is for food production.

circulated a petition to transform a golf course in Minneapolis that sits slightly lower than the adjacent Lake Hiawatha into an extensive food forest mixed with recreational areas.[15] The goal would be to address major water issues related to both pollution and flooding and incorporate the Dakota people's historical wild rice harvests into the experience.[16] *Designing Urban Agriculture*, by April Philips, provides an excellent overview of similar projects from around the world that retrofit cities to better support sustainable urban agriculture.[17]

The authors of *Carrot City: Creating Places for Urban Agriculture* pose two important questions that are relevant to community food forests.[18] First, how can design reflect emerging movements that encourage us to consider ourselves coproducers rather than consumers and engage in food production and supply processes? Second, how are underutilized urban spaces being transformed by current food production projects, and what further untapped opportunities for food production exist?[19] Urban agriculturists can help community food forest leaders answer these questions, particularly through their expertise in designing food production areas in underutilized spaces throughout a city.

In *Breaking Through Concrete,* writer David Hanson and urban farmer Edward Marty define urban farming as "an intentional effort by an individual or a community to grow its capacity for self-sufficiency and well-being through the cultivation of plants and/or animals."[20] Community food forests meet this definition because they contribute to local food production and community stabilization. Project leaders should closely consider how to balance annual and perennial production in relation to the needs of a community. If food security is the main concern, food forests should be carefully designed for high yields in both the short- and long-term. If production quantity is paramount, it may be best to use an agroforestry approach where perennial and annual species are mixed together using any number of practical polyculture patches or systematic row plantings with short-term annual production of vegetables or small livestock if codes allow. Private urban agriculture business models could be followed to create a public-private arrangement such as a CSA or direct-to-consumer sales.

URBAN GARDENERS

Formal community gardens have clearly played an important role in shaping our urban land use legacy. They provide multiple benefits to individuals and communities, and there is much to gain from working with urban gardeners on community food forest projects. Many lessons learned in developing a community garden can prove useful in creating a sustainable and successful community food forest.

The community garden movement has grown by leaps and bounds in the past twenty years and is now one of the most common forms of urban agriculture. Unlike the era when community gardeners were associated with wartime austerity and economic hardship, community gardens today are a visible and permanent part of many communities. One of the country's model contemporary community garden programs began in Seattle in the 1970s under the name P-Patch. Today the city lists community gardening in its urban development plan, and the Beacon Food Forest project in Seattle (described in detail in chapter 13) partnered with the P-Patch program to secure the city's consent for the food forest on public property.

Gardening is often a key component of community food forests. Perennial production is a selling point for most projects, but some aspect of annual gardening is usually also included, whether as independent plots or part of a dedicated program such as youth education or therapeutic gardening. Lastly, gardening is generally a highly prized activity that provides a diversion from the daily grind. Gardens also reflect our cultural values and personal beliefs. They are places where people go to work, heal, and learn. They provide us with a direct means to a restorative relationship with nature. As such, gardening opportunities help to sustain overall interest in a community food forest among volunteers and visitors.

Concerns about public gardening tend to revolve around cost, liability, and maintenance. Community food forests share similar issues. Much of what is changing in support of urban gardening also improves the political and cultural conditions for community food forests. Both seek to increase efficiencies in the use of urban space with a focus on food production and stronger connections to the natural world. Children who garden generally have a better understanding of nutrition and eat healthier, which is also a common goal of community food forests. The cost and availability of fruits and vegetables inhibit availability for many. By increasing access in an affordable way, community gardens help increase consumption, as do community food forests. Other shared benefits include opportunities to connect with community members through food production in a social setting.

Producing food together and connecting with nature in a structured and supportive place such as a community food forest offers people a safe and comfortable civic space similar to a community garden. Community is built and strengthened through relationships that begin and grow when people interact, work together, and share at the site, which can be a combination of a garden and a food forest. For example, we installed a food forest along the edge of the Hale-Y community garden in Blacksburg, Virginia, which opened up opportunities for engaging socially across a larger network of people. Gardeners learned from community food forest leaders and vice versa, and all developed a neighborly assistance network that grew out of exchanges such as trading strawberries from the food forest for lettuce, cucumbers, and other vegetables from the gardens. One gardener brought lemongrass to the food forest to expand production, and others were able to learn more about perennial species, management methods, and products. The food forest became a place of conversation about managing forests for products, aesthetics, and pollinators. Activity spilled over from one to the other, such as a session picking fruit after weeding garden plots. A social and ecological polyculture was created, and all involved strengthened their role in local food production and personal investment in the community as well as the environment.

URBAN WILDLIFE SPECIALISTS

Wooded habitat typically houses a diverse array of animals, and planning for wildlife is often a key consideration for farmers and gardeners because people want to protect crops. Bringing edible crops into the urban landscape introduces additional liability, especially in terms of protecting children and the vulnerable, but also because wildlife will be attracted to the site. Thus, planning and stewardship are needed to ensure rightful management for all that a food forest impacts and benefits. It is prudent to study potential wildlife visitors, unwanted and wanted, and to seek planning, design, and management guidance from wildlife scientists.

Some of the community food forest projects we visited looked closely at this during the design phase, working with wildlife specialists from the outset. Patterns that support preferred microfauna such as pollinators and soil biota were identified and installed, along with those that help keep out potentially destructive macrofauna such as bears. Other community food forest groups have learned the hard way much later in the process. We recommend studying potential wildlife management by working with specialists in your region, be they employed in the private or public sector.

Many projects have experienced the wonder and worry of wildlife as they evolve and expand, and there are numerous examples of community food forests that have creatively merged humans and wildlife. In the Roger Williams Edible Park in Providence, Rhode Island, for instance, the food forest is alongside a large body of water. It came as a pleasant surprise when project leaders found turtles nesting in the site; the area was marked off and signs directed users elsewhere. By emulating a forest ecosystem, and in this case a riparian woodland buffer, the designers had created habitat that was then occupied by wildlife that prefers the setting. If something similar happens in your community food forest, take steps to ensure mutually safe environments for both human and wildlife visitors.

The appearance of wildlife can be a sign of success at a community food forest, but it can also be a signal that management actions are needed to protect both the fauna and unsuspecting community visitors. Grant funding may be available for creating habitat to support pollinators or species in decline, and securing such funding can have an enormous impact on project feasibility. At the Hazelwood Edible Forest in Pittsburgh, Pennsylvania, milkweed began to grow, and the decision was made to leave it as a food source for monarch butterflies frequenting the site. Milkweed was not one of the original species in the design, but because it appeared naturally, the food forest team decided to put it to work and learned how to make pickled milkweed pods. They were also able to share their story about monarch habitat conservation when

FIGURE 9.5. Some forms of wildlife have a bad reputation but are, in fact, beneficial in gardens and food forests. Arachnids (spiders), like this common yellow garden spider (*Argiope aurantia*) at a community food forest in Syracuse, New York, may be a little creepy but are not dangerous to humans and help control insect populations.

FIGURE 9.6. Food forests can provide valuable wildlife habitat. In the Hazelwood Food Forest in Pittsburgh, Pennsylvania, this monarch butterfly is resting on its primary food source, the milkweed plant. Purple flowers attract a number of other butterflies and moths.

serving the pickled pods to others. Another community food forest is a pollinator paradise with species such as borage, yarrow, catmint, bee balm, lemon balm, lavender, echinacea, and other flowering herbs. This particular food forest buzzed all season long with insect life, including multiple butterfly species, elevating visitors' sensory experience.

Core leaders at the Basalt Food Park in Basalt, Colorado (described in chapter 10), worked closely with a wildlife scientist to safeguard its community food forest from large animals, such as elk and bear. At the Beacon Food Forest in Seattle, honey production was part of the plan, but the beehives were carefully placed away from main walking areas. Two other community food forests we visited included multiple bee houses hung throughout the food forest to attract native bees and to give carpenter bees places to nest rather than causing damage to infrastructure.

CULTIVATE KANSAS CITY

Cultivate Kansas City integrated a food forest into their agricultural business model, but the organization also offered it for use as a community educational asset. The organization was founded by Katherine Kelly and Daniel Dermitzel and was originally known as the Kansas City Center for Urban Agriculture. Katherine and Daniel connected as allies through their shared urban agriculture interests and networks. Katherine had a long history in backyard and community gardening, worked as a biodynamic farmer, and is the current executive director of Cultivate Kansas City. Like Katherine, Daniel was an organic farmer with experience in urban planning and communications. He was the associate director of Cultivate Kansas City and an adjunct professor in the department of architecture, urban planning, and design at the University of Missouri in Kansas City.

Daniel and Katherine met and formed the idea for Cultivate Kansas City in 2005, which involved converting Katherine's Full Circle Farm into a demonstration and teaching center. In 2008, Daniel completed a permaculture design course at the Central Rocky Mountain Permaculture Institute, in Basalt, Colorado. He joined the Kansas Permaculture Collaborative to continue learning about and engaging in permaculture design. Cultivate Kansas City began by promoting neighborhood food production, providing public education on urban agriculture and related issues, and working with independent farm businesses in the area.

As the local food system around them changed, their mission and scope of work expanded. By 2010, they were advocating for policy issues among local government agencies and the city council to update and add new codes in support of urban agriculture. They also became a major player in the Food Policy Coalition of Greater Kansas City and began a CSA program, offered food production and water access mini-grants to farmers and community gardens, and provided other educational and technical resources. Educating the public and training beginning farmers are two main goals of Cultivate Kansas City. In collaboration with other local organizations, institutions, and nonprofits, they host summer farmer apprenticeships at the demonstration farm and the food forest.

As Cultivate Kansas City grew, it was decided that the group should focus on their main two-acre demonstration farm, Gibbs Road Farm. This narrowing meant that one of their lots located in a nearby neighborhood was taken out of food production. It was a quarter-acre corner parcel that had a history of agricultural use

FIGURE 9.7. The Cultivate Kansas City plot was used for urban agriculture for decades. Now planted with perennials, it is transforming into an early successional forest with fruit and nut trees. This site was designed more with production in mind than creating a public space with clear pathways, infrastructure, or defined gathering spaces. But once one enters and explores, there are diverse foods to be found like these cherries.

since the early 1900s, first as a dairy farm, then a market garden, and finally a personal garden. A volunteer couple, PJ Quell and Larry Davis, owned and managed the lot in cooperation with the organization. In 2010, the organization decided to plant the site in cover crops to buy time while deciding what to do next. When a board member discovered a grant opportunity to further develop the site, the discussion turned to the role of a food forest. The grant was called Together Green and was jointly funded by a partnership between Toyota and the Audubon Society to support creative projects that engage diverse communities in measurable conservation.

Daniel applied for the grant to do research on food forests and learn more about production possibilities. They received a $10,000 grant and ultimately decided the best way to learn would be firsthand by installing a food forest on the corner lot. Daniel had formed a strong bond with PJ and Larry, and they agreed to plant their land with perennial species. He sought to design the food forest so that it would serve as a legacy project aligned with their goals, providing a place they could enjoy, while also offering a perennial agriculture demonstration others could learn from.

Three swales were created on the site in late 2010 and the lot was seeded with clover. Through the Kansas Permaculture Collaborative, Daniel consulted with professionals with a deep understanding of perennials to improve project

> **Perennial Practice**
>
> For many, farming or gardening is a spiritual practice, and this surfaced in some of our discussions with food forest leaders. Several community food forests are located on the grounds of religious institutions seeking to gather people, feed them, and contribute to spiritual community by working the land together. But regardless of where a community food forest is located, people have expressed a sense of solace or stewardship by connecting with the earth and providing beneficial services to each other. All benefit the human experience and are ways of growing human capital.
>
> People in Kansas City were drawn to Cultivate Kansas City's food forest project because its principles reflect a vision that is sacred and resonant. The food forest is positioned as a way to give back to the earth and create something tangible and nourishing. Through sharing the harvest, it is also a symbolic act arcing toward the humanity people aspire to in their communities and beyond. This essence reflects the idea of a food forest as a communal perennial regenerative form of agriculture and source of human capital that when amplified in social terms forms a higher order of cultural capital. This potential for community food forests to serve a community and individual well-being is worth reflecting on when developing project and site design goals and objectives.

design. He collaborated with horticulturalists and orchard owners to help choose appropriate species and explore what other plants and trees could be grown in the area. The first phase of the food forest was planted in spring 2011. It was mostly a canopy layer with varieties of apples, cherries, serviceberry, persimmon, pear, flowering quince, Chinese chestnut, jujube, peaches, pawpaw, plum, elderberry, goumi, flowering pea shrub, sea buckthorn, linden, and New Jersey tea. A year later, they followed with raspberries, aronia (chokeberry), blackberry, blueberry, goji berries, and currants. The herbaceous layer was installed in 2013 using species that the Gibbs Road Farm could sell at the market. More was added in 2014. Everything was selected for its potential market value and sales to local restaurants.

After planting the second layer of vegetation in 2012, Daniel had the opportunity to spend a year designing and installing organic farms, called the Happy Farms, at Thich Nhat Hanh's meditation center, Plum Village Monastery, in Thénac, France. Farms there were created to actively promote mindfulness, community building, and a deep understanding of sustainability by having those who stayed at the monastery for studies or retreats participate in the life cycles of the farm. Organic farming was considered not only a sustainable practice at the monastery but also one of mindfulness.

When Daniel returned, he shared what he learned about farming as a mindfulness practice with the food forest team. As a result, Cultivate Kansas City began to shift its sense of sustainable agriculture. While production methods on the demonstration farm were ecologically sustainable, socially they were not. The constant demands of annual production across distributed parcels was taxing on the people involved. With this in mind, Cultivate Kansas City began incorporating trees into their farms, growing herbs in polycultures, focusing more on longer-term visions, and becoming mindful of balancing time and expectations.

Apart from shifting how Cultivate Kansas City practiced sustainable agriculture, the food forest rendered other key benefits. It connected the organization's members to new circles of growers, educators, and community groups who eventually became involved in the site either through workshops or activities such as happy hour, yoga, or morning coffee. It also led to new products to sell at the market or to local restaurants. Leaders learned about the challenges and perennial food production successes and were able to pass on lessons learned to other beginning farmers.

A community food forest designed for urban agriculture will look very different from one designed for public space unless special attention has been given to include design elements of both. The Cultivate Kansas City food forest was not designed to be a public space, but rather as an alternative urban agriculture site for production and hands-on learning. Although many gained a spiritual sense of peace through working with the site and the perennial cycles of growth, few people from the surrounding neighborhood were observed visiting the site.

From this perspective, some changes could help transform the food forest if the goal were set to increase use among neighborhood residents. The swales are doing their job to collect water and support production, but the result is barely visible and wet pathways remain during the summer months, which may discourage exploration by local residents. There is no infrastructure or defined space for sitting to reflect or rest within the food forest. Instead, reflection and mindfulness are achieved through action and repetition, but that is not possible for those who have difficulty walking or are beset by other physical challenges. Adding these elements could increase curiosity among passersby, but in this case, this is not an explicit goal.

In fact, the space contributed to community in a different way—it was used as a research site by Stephen Mann, an agroforestry graduate student at the University of Missouri. Stephen was involved with the site through permaculture activities and started recording data there in 2014 to study the biogeochemical, ecological, agronomic, and economic parameters of a temperate food forest. In 2015, the site was included in an annual urban farm and garden tour and research information was added to the site's educational experience programs.

Depending on the specific goals and intended uses of a community food forest space, there will need to be a carefully designed balance between agricultural production and its role as public space. In this chapter, we provided examples of allies that can help achieve that balance, but ultimately the final design is up to the community. It could be parklike or production-oriented. A potential solution is that specific areas in the design can be dedicated to different objectives to accommodate the multiple benefits a community food forest is intended to provide.

CHAPTER 10

The Basalt Food Park

In this chapter, we share the Basalt Food Park story to demonstrate the power of community food forest teamwork through successful publicity and partnerships with municipalities. It also showcases the importance of perseverance among core project leaders and the role of earth-based businesses and agroecological experience in creating a diverse project. The teamwork and commitment among those involved with the Basalt Food Park provides a model snapshot of how people come to participate in a community food forest—and how to capture, channel, and amplify their motivation and energy. It is a parable of the power found when energy is focused on accomplishing goals rather than succumbing to the ups and downs of a project.

The benefits of partnerships with municipalities are also on display in this chapter. Project leaders worked with their town officials and other local natural resource technical experts to address challenges and find answers to important questions about how to pinpoint important environmental and social benchmarks, mechanics, and assumptions that lead to strategic success. As you read through the narrative, take it in and think about the aspects of planning you and others have already addressed for your project and try to identify aspects you have yet to tackle.

Fast Facts

PROPERTY OWNER: **Basalt Recreation Department**
LOCATION: **Ponderosa Park, Basalt Ave, Basalt, CO 81621**
YEAR PLANNING STARTED: **2013**
YEAR ESTABLISHED: **2014**
SIZE: **0.5 acres**

ABOUT BASALT

Basalt, Colorado, sits in the Roaring Fork Valley and was incorporated as a railroad town in 1901 by homesteaders who laid claim to the area for coal mining. Today, it is midway between two tourist destinations, Glenwood Springs and Aspen. Glenwood Springs is a popular spot known for its hot springs and spas, and Aspen is recognized for world-class skiing. Basalt has a downtown area of approximately two square miles and had a population in 2013 of less than 4,000. The economy is supported by tourism and second-home owners who are attracted to the quaint town surrounded by mountains and situated at the doorstep of a vast expanse of amazing Colorado landscape.

Basalt also sits at the convergence of the Frying Pan and Roaring Fork Rivers, which makes it an attractive fishing location. Under a rails-to-trails program, the old tracks running through the town were turned into the Rio Grande Trail, a forty-two-mile protected multiuse corridor that offers biking and hiking between Glenwood Springs and Aspen. The Aspen Global Change Institute, the Rocky Mountain Institute, and the Central Rocky Mountain Permaculture Institute are located in Basalt and contribute to its environmentally conscious and sustainability-oriented community.

In the early 2000s, Basalt was primarily a retirement area, but like the rest of the Roaring Fork Valley, it has experienced noticeable growth and a shift in demographics toward younger, diverse families. Today, the Hispanic population now makes up around 20 percent of the overall town population and 60 percent of the primary and secondary student bodies. As Basalt grows and changes, the community has asked itself serious questions about how to maintain a small-town feel in light of its growth and evolution. To answer these questions, citizens have turned to green infrastructure and have supported environmental development such as parks, open spaces, and trails systems that connect parts of the community, expand recreational opportunities, and minimize the need to travel by car. Since 2006, Basalt has levied a 1 percent sales tax to generate revenue exclusively to support such community-specific green projects.

IDEA GENERATION

Stephanie Syson moved to Basalt from Florida to manage Colorado Rocky Mountain Permaculture Institute's (CRMPI) greenhouse and education programs developed by founder Jerome Osentowski. CRMPI is located on Jerome's residential property and is known for its diverse and productive food forest, which is split between a large greenhouse and an outdoor area complete with a pond and chicken coop. The dry climate, alkaline soils, and short growing season in Basalt were a major change from Florida's verdant landscape where plants grow easily and quickly. However, Stephanie's work at CRMPI helped her gain a new perspective on the range of possibilities for cultivation in Basalt's climate. She also began to realize two problems faced by residents of her new town. One problem was that the vegetable, flower, and herb seeds available through mainstream companies are typically not acclimated to Basalt's environment and often fail to germinate or thrive once growing, which discourages novice gardeners. The second problem was that food supply can be tricky when heavy snowfall or avalanches restrict access to larger cities in the north or south. These two problems spurred Stephanie to develop a seed library project with the local public library, which included both a seed distribution program and an accompanying garden. As you will learn, what started as a library project soon evolved into the redesign of a local park from little more than open green space with a path into a re-enlivened place with a community food forest that includes a seed garden, outdoor musical instruments, seating areas, and educational opportunities.

BUILDING SUPPORT AND FORMING PARTNERSHIPS

In the wake of the digital age's downloadable books and free online content, libraries are searching for ways to stay relevant and draw people to their public space. When Stephanie

THE BASALT FOOD PARK

FIGURE 10.1. The seed bank in the Basalt Food Park is well marked by signs and various plants such as this lettuce maturing to the flower and seed stage. Seeds are harvested by volunteers and given to the library, where they are cataloged according to the place they were collected. They are then divided into small packets for patrons to check out and plant in their home garden.

spoke with the Basalt Public Library about creating the seed library project to conserve genetic diversity and improve local access to viable stock, the idea was welcomed. The seed project and its mission to provide seeds and plants to the public seemed like a perfect addition to the library. She tapped into another network of local food system advocates, the Roaring Fork Food Policy Council (which has been renamed the Roaring Fork Food Alliance), and formed a partnership between the council, the library, and CRMPI. Together they were able to solicit seeds from eight local sources to start the project.

With Stephanie leading, they held workshops to introduce the project to the public.

The primary aim of the Basalt Seed Library Project is to help the public develop gardening and seed-banking skills. Seeds are placed into small envelopes similar to the packets found in stores and categorized according to seed-saving difficulty to make it easier for patrons to determine what they would like to try to grow. Library patrons are allowed to take up to ten seed packets from the library with the understanding that the healthiest and most viable plants growing from the batch will serve as a source of seeds that are to be returned to the library for future distribution. In order to keep costs low, the seed project was established on land where vegetables, herbs, and flowers could be maintained for future seed supply to stock the library program and give the community time to establish a cycle of a lending and returning. The project is part of larger network of seed libraries throughout the country, many of which operate out of public libraries.

As the project developed, it gained attention and was publicized through local media and then covered in a national NPR story. Lisa DiNardo, Basalt's horticulturist and arborist at the time, heard the story on the radio and loved the idea. She called Stephanie to introduce herself and offered to help with the project. They quickly realized they shared interests and another partnership was formed. Lisa identified public land close to the library that could be established with seed project plants to supplement the seeds harvested from personal gardens around town.

Among the public land Lisa identified, Ponderosa Park, a large stretch of underutilized land in a highly visible section of town, was selected as the best option. Located across the street from a shopping center, the park is also adjacent to a low-income neighborhood with a high percentage of Hispanic residents, and along the primary footpath between the main highway (Highway 82) and a footbridge over the Roaring Fork River to historic downtown Basalt. Another shopping area and the Basalt elementary and middle school are located directly across the bridge, creating a continuous flow of heavy foot traffic throughout the day.

Adding Up the Capitals

Basalt Food Park leaders put human capital (experience, knowledge, and energy) to work as a means of investing in locally grown Basalt food and its gardens and gardener culture. They amplified the reach of their skill sets by tapping into relationships built through professional connections and interests. Lisa, who worked in a town agency, had access to the city's equipment and resources. She was in a good position to communicate with town staff about changing a land easement on the park to help ensure that the food forest could be established. By virtue of teaching courses in the local school system, Stephanie was connected to educators and was able to quickly make use of the site and show its educational utility. Both women sought to work with the public library and its grounds to make the project a reality. They invested their energy and connections to prepare the site and to retain the commitment of volunteers after the first workday. Volunteers enhanced their personal skills and grew community relationships. Partnerships formed through the seed-saving project and multiple community capital investments added critical human, political, and financial assets that helped the garden evolve into a food forest and enhanced environmental services for the town.

INSTALLATION AND COMMUNITY ENGAGEMENT

Stephanie believed Ponderosa Park was well suited to be an educational demonstration site and wanted inclusivity to be a reality from the beginning. She invested in banners and signs to inform community members about the project, printing them in English and Spanish. During volunteer workdays, Stephanie arranged for a local bilingual community member to provide translation services and to invite passersby to get involved. Support for the project was overwhelming, but as is often the case, there were unique and unexpected challenges along the way.

Deer, elk, and bear are problems in the Basalt area, and there was concern a community food forest would attract large game. Fencing around the food forest to keep out animals was needed to address concerns of the Colorado Division of Parks and Wildlife as well as local residents. Matt Yamashita, the district wildlife manager, worked with Stephanie and Lisa to design an exclosure that would keep large wildlife out. However, regulations posed an obstacle: The proposed plot in Ponderosa Park was next to the river and under conservation easement. The easement disallowed fences, but Lisa was able to work with the former owner to revise the easement so that it would allow for installation of fences and some benches. She worked with officials in the town planning department to produce a properly worded document.

Lisa and Stephanie intended to start small by planting the seed-saving garden on a

6,000-square-foot plot near the most visible location within the park. They purchased fencing for the area and were set to begin. However, plans changed as the town identified a potential financial backer who was constructing a state-of-the-art "green" innovation center nearby. One of the sustainability goals of the project was to establish 11,000 square feet of urban farming to offset the center's environmental footprint. Developers of the innovation center became interested in the Basalt Food Park because of its location near schools and community retail centers, and after discussing the possibility and benefits, Lisa and Stephanie agreed to add 5,000 square feet to the seed-saving garden to fulfill the 11,000-square-foot requirement. In light of the added space, it made most sense to install a fence around the entire park, but there were concerns it would denote a community garden, where plots are allotted for individual cultivation and access is limited. But Stephanie hoped that news would spread by word-of-mouth that access was open and that the garden was a new model where anyone could take part in creating, maintaining, and sharing food.

They launched the project using a specific type of fencing that is required by the Colorado Division of Parks and Wildlife. However, they planted only the initial 6,000 square feet of the seed-saving garden while they waited for additional information on funding. In the end, the potential funder became entangled in a long process of approval for the innovation center and some parts of the seed-saving project did not fit within the funder's standards. Eventually, the financial support fell through.

A pattern across community food forest projects is that when tangible financial assets fall through or are delayed, intangible assets such as human and social capital are the resources that carry a project forward. With help from Rifle Correctional Center inmates, a minimum-security prison that contributes labor to community projects, Stephanie and Lisa were able to install the fence and optimism remained that the food forest would still take shape, which in turn kept them excited about the project. Lisa continued to communicate with the town of

Knowing Your Community

Demographic information can help inform infrastructure planning at your community food forest. A comprehensive picture of who lives nearby can help create a design that ensures equal access for people of all ages and uses plants that provide produce for a variety of diets and preferences. It can help determine the signs and information that meet the needs of community members and build support for the project.

A strong connection to community values utilizes cultural capital to attract social support from the immediate community, but also from visitors. Identifying area businesses or organizations that have similar values can pay off as sources of political and economic capital, which contributes to the food forest's infrastructure and environmental services. In Basalt, for example, the leaders knew the adjacent neighborhood included many Spanish speakers, so they invested in having a translator present at events and had signs printed in both English and Spanish.

FIGURE 10.2. The Basalt Food Park in its infancy in 2014. The freshly planted site may not resemble a young woodland yet, but plants like comfrey will expand as ground cover and create organic matter that helps improve the soil. The fruit trees were strategically placed far apart to minimize competition for sunlight and nutrients. Berry bushes such as the red currants will create a middle layer in the understory. Here, they have been placed away from the fruit trees to minimize competition for light.

Basalt, and Stephanie was able to build excitement for the project through her work with the public library and local schools. When an opinion piece in the local newspaper branded Stephanie as a utopian communist, she stayed positive, responding "How else do you make change? You have to give it a whirl."

This is an example of people in the inner circle sticking with it and shepherding attitudes that are positive, optimistic, and patient, which can strongly impact project success. Jimmy Dula was another core group member and the founder and owner of Colorado Soil Systems, a company that educates valley residents about fungal processes and healthy soil. Jimmy was an enormously important asset during the establishment phase because of his "get-it-done" attitude, which helped put the installation into motion. When we visited the Basalt Food Park, Jimmy was well invested in community food forests and getting ready to start another project in the nearby town of Carbondale. Now that the community food forest is installed in the Basalt Food Park and receiving positive coverage by multiple media sources, the power of their positive attitude is validated.

Through the process of preparing the park for installation of the fence, seed garden, and food forest, the team learned some important lessons. When working with public property, multiple stakeholders should be at the table before project plans move forward to prevent pushback or roadblocks once it starts. However, coordinating schedules for meetings can take time because the food forest is certainly only one project among many that local officials and others are involved in. A flexible timeline and deadline modifications were necessary to develop a food forest with all the right measures in place. Planning for just the food forest component of the Basalt Food Park project took place through multiple meetings over the course of a year. It was apparent that solidarity and support of dedicated partners in the struggle are crucial when things seem untenable.

Project vision was also essential to maintain motivation. Lisa and Stephanie's shared vision gave them something to work toward. They expressed a spiritual connection to the project in that it helps fulfill larger planetary needs at a small scale and provides people with a place where they can work and care for a greater good. This sense of acting on behalf of humanity was a sustaining force when facing project challenges. In addition to a shared vision, they found it was helpful that one of them worked within town offices. As the town horticulturist, Lisa was in a position where she could hear what other town staff thought about the project and address concerns as they arose. She was also able to keep the project at the top of the list of priorities. At a deeper level, she was also able to share the project vision with her colleagues.

After more than a year of planning, the food forest installation began in spring 2014 with the help of volunteers. The partnership with the town of Basalt was helpful in that mulch was provided free of charge and all necessary equipment was available on loan to complete site preparation. In anticipation of the first volunteer planting day, Lisa arranged to have town-owned heavy equipment dig deep holes on the site where large trees were to be planted. Jimmy, Lisa, Stephanie, and a few others prepared the biggest holes and trimmed pathways in the park so that on planting day volunteers would not be overworked. Instead, the volunteers could focus on digging small holes, sowing and planting, and mulching the site. Overall, providing a fun and exciting experience was an important goal of the leadership team, so all who participated would enjoy a sense of accomplishment that translates into motivation and support.

CONTINUATION AND FUTURE VISION

The flexibility, patience, and positive attitudes paid off in spring 2014 when the previous year of planning transformed Ponderosa Park into the Basalt Food Park. The revised and expanded installation had three distinct sections: the native woodland, the seed library, and the food forest. Purchasing older trees to install helped quickly establish a noticeable young forest. The seed library is located near the front gate where it attracts attention and draws people into a section with familiar annual plants before they walk through the native woodland to the community food forest. The fence became a functional billboard, with educational signs in Spanish and English attached along its length.

Less than one year after establishment, the Basalt Food Park began hosting educational tours and school programs that teach students about food production, seed banks, and ecology. By 2015, the focus began to shift from the work of planting to developing additional signs and hosting even more educational school field trips. Support among the community as well as the town government began to increase as well, resulting in positive feedback regarding the project's community investment.

Jimmy hopes people will use the food forest to learn about how they can improve soil health where they live. He added compost and worm castings at the food forest to encourage microbial, bacterial, and fungal growth that enrich soils. The town deposits mulch at the site to help keep moisture levels higher in the dry climate, and Jimmy sprays a compost tea on the layers of soil and mulch to add nutrients. When someone takes a plant from the food forest and adds it to their home garden, they can also collect soil and mulch to help with transition and establishment where they plant.

The Basalt Food Park site provides family opportunities as well. Children can play king or queen of the mulch pile and pick berries to eat. People can stop on their way home to sit and listen to the Roaring Fork River nearby and observe bird and insect life. One extraordinary feature of the site is a music garden that includes weather-resistant musical instruments anchored to the ground. The instruments attract a lot of attention, and when people stop to play one, they face in the direction of the food forest, where signs have been strategically placed along the fence that describe the garden and where one can enter. There is interest in adding more artistic installations to the site in the future to attract more people. One suggestion is to establish a grove of aspen logs inoculated with oyster mushroom spawn that would stand at various heights in creative patterns.

With the addition of a community food forest, the park became a multifunctional resource that the town holds up as a useful part of Basalt's effort to enhance community experience, protect its way of life, benefit the environment, and provide a place for community members to relax and eat! All of this is possible in terms of the park's overall vision to increase plant accessibility for Basalt residents and make gardening easier, cheaper, and underpinned by regionally acclimated stock. The hope is that increased access to appropriate planting material will help people plant them on their own property or coordinate planting in common spaces near their home.

Another goal is to increase structures in the park so that people can use the space for gatherings and events. Future plans include installing a wood-fired pizza oven, picnic tables, trash receptacles, and more seating. Ultimately, all the vegetation and infrastructure installations are turning a park that was once mostly bare and used as a pathway into a hub for education, play, relaxation, gatherings, celebrations, and a resource for spreading seed throughout Basalt.

PART 4
Purposeful Community

CHAPTER 11

Reflecting on Community

Critical reflection is not a simple musing on project motivations or an academic exercise for pondering vision and goals. It is a process that deeply examines the meaning of the experience being created. It requires asking questions that are not easily or quickly answered. When practicing critical reflection, we must admit where we are falling short. The process also illuminates where we need to continue learning, which ultimately shapes better leaders. In this chapter, we present a few methods for critically reflecting on community food forests and the process of project development.

Critical reflection is important at several stages when creating a community food forest. One instance is when interest in the community food forest begins to form and potential locations are identified. It is important to note that the interests and views of people who live in the place where the community food forest will be located are not always the same as the interests and views of those who want to create it. Oftentimes they are quite different. However, community in some form is what typically emerges.

Another time to reflect is after the project has been functioning for more than a year, by which time regular volunteer commitment has formed, and hopefully, a community will have materialized around the food forest to support its growth. The question is, who is in that community? Understanding the differences between outside interests and local dynamics and needs is important, because the divide can be either a sticking point that ends a project or a springboard that inspires a functionally diverse partnership.

BONDING AND BRIDGING CAPITAL

As we explained in chapter 3, bonding and bridging are two forms of social capital that are helpful for identifying ways to design a project that clearly cultivates community support. Bonding social capital tends to be the form most easily developed during planning for a community food forest project. People with similar interests and values are generally willing and eager to work together and thus constitute a formative community. Bridging capital typically comes a little bit later, when there is a need for networking across heterogeneous groups that may be involved in or affected by the community food forest.

Bloomington Community Orchard in Bloomington, Indiana, went through growing pains during its first year as the core group struggled to understand how best to move

FIGURE 11.1. Volunteers join hands at Seattle's Beacon Food Forest before a work party to solidify teamwork. Small rituals and extra effort to bring people together in unity creates social bonding. Photo courtesy of Jonathan H. Lee.

their project forward, quickly apply for and win startup grants, and put in place a consent-based management process that helped them learn as they went along. (We provide detail about this community food forest in chapter 15.) Project formation was quite painful at times, yet without the struggle and local support, the team would not have formed the familial bond they now feel or the deep commitment to work together to realize each other's goals. The point here is that bonding and bridging capital do not occur only during joyous celebration, but often more so through shared struggle and mutual support as community food forest progress ebbs and flows.

Bridging capital is typically harder to build because it entails working across different groups, some of whom may not see eye to eye or share interests and perspectives on what a community food forest can and should be. It is nevertheless important if a goal is to include residents from ethnically diverse, low-income, or high-crime neighborhoods. Project leaders at more than one site we visited raised the issue that team leaders are mostly middle class and white, though stated aims of most community food forests are to be as inclusive as possible and build participation across a wide demographic. These leaders expressed a need to include others as well as sensitive self-awareness regarding their lack of knowledge or experience effectively engaging diverse groups. As such, bridging capital takes time. Often the best place to start is by recognizing that it is lacking, motivating a core team to pursue it, and accepting that investments will take time and require patience.

Keenly interested community food forest stakeholders are the easiest to identify because they are the people who step up early and join the cause. They identify with and share the values and concept of the food forest regardless of where they live relative to the site. They are often the folks who attend initial meetings and regularly take part in planning and advocacy. Their personal and social identity may be tied to their main interest, whether that be as a foodie, permaculturist, social activist, political anarchist, local food system advocate, or conservationist. The feeling of satisfaction from engaging in an activity that fulfills the image of one's identity in relation to their passion is a profound benefit. It can be a major incentive for continued involvement in a community food forest either as a volunteer or leader. In short, these motivated advocates are very approachable about community food forest projects and usually bond fairly easily with other volunteers. This inclination to join is a powerful unifier and can lead to the rapid buildup of collective strengths across diverse interests. However, if bonding does not include local residents, it can also be a limiting factor for perspective on the scope of the project and how, if at all, it can best serve community members near the site.

Bonding social capital also creates cultural capital. Bonding grows cultural capital when people in a certain community form and share similar tastes in style, ways of doing and thinking about things, or language associated with a particular interest or hobby. This strengthens collective identity, which might alienate those who are not considered to possess the same type of cultural capital, or who see themselves as not having enough to be a part of a group. The relationship between bonding and culture is an important consideration when engaging community members in terms of bridging strategies.

THE MEANING OF COMMUNITY

People who live in close proximity to a community food forest may or may not be interested in it. A different type of outreach may be necessary to connect with them compared to those who involve themselves out of interest in community food forests alone. Many of the individuals, families, and civic leaders in proximity to a community food forest have a different relationship with the site because they witness the day-to-day activity in the area and are usually more attuned to a broader array of community issues and needs. Moreover, they may already use the site in other ways or have different thoughts about how to develop it. The majority of the sites we visited were located near or in low-income, high-crime, or ethnically diverse neighborhoods.

The fact that many community food forests were located in these areas may be the result of various factors, including a desire to offer local access to nutritional food, the availability of vacant lots or marginalized land, or an increased chance to garner grant funding. Thus, something seemingly as simple as site selection should be the subject of critical reflection. It cannot be only about the environment or saving a troubled neighborhood. Fine-scale considerations about the human dimensions of a site are also warranted.

Asking hard questions about alignment between external interests and local needs at the beginning of a project, such as those listed below, can help develop a truthful map of the community where a food forest may be located.

In doing so, it can also minimize the challenges of intersections between people who are interested in creating change through a community food forest and those who live in the places where these projects may be installed. A merger of both during project formation can head off conflict and build an inclusive social system from inception. The following questions are examples of inquiries about bridging and bonding at community food forest sites:

- Is establishing a community food forest a socially acceptable and reasonable way to build community, promote healthy lifestyles, contribute to local food systems, and rehabilitate the land?
- Is it possible to reach out to local community members or representatives first to determine needs and willingness to engage in a community food forest project?
- Will a community food forest be seen as a product of dominance, power, control, or privilege among local residents and others in the project area?
- If one concern is about addressing a food desert (an impoverished neighborhood with limited access to fresh and wholesome food), what types of outreach can bring different interests together, rather than one "fixing" a problem for the other?

An analogy using permaculture practices can provide a useful strategy for reaching out to and working with diverse stakeholders. Consider how permaculture requires site analysis and observation before completing the ecological design of a food forest. The scope is comprehensive and includes factors such as climate, terrain, energy passing through the site (as light, water, or wind), and other characteristics that might affect selected species and preferred management practices. The same depth of analysis is important when observing and analyzing a neighborhood before organizing a community associated with a food forest.

Assessing community capitals, history, and relationships between various groups that are involved with or potentially impacted by a community food forest can shed light on how the political and economic climate, social and cultural terrain, forces of influence, and other characteristics drive what is more likely to work at the site and who will be involved. For example, a group may state a goal to keep the site open to the public, but a variety of barriers and potential pitfalls must be considered, including social structure. Even though a site is not fenced, for instance, some people may view it as enclosed or off-limits. Such social or cultural barriers are far harder to cross over or to remove than something physical. It is not a cultural norm in society to harvest food freely from a public place or from private land tended by someone we do not know. It is even less likely when differences in status are at play.

Such deeply entrenched social patterns are difficult to change or replace. Another useful strategy for managing sensitivities to social patterns is critical discourse analysis, which, like critical reflection, involves reviewing what is being said about a project and analyzing it and carefully judging its merits and faults. The explanation is that the language we use influences how social relationships are reproduced or challenged. It considers such things as whether repeated motivations and goals for creating a food forest become a mantra in selling the idea to an elite class or are crafted in a way that includes disparate voices in dialogue rather than silencing them.

MEETING COMMUNITY NEEDS

Collecting diverse ideas and designing a site to be multifunctional helps cultivate resilience and long-term viability. One of the first questions should always be whether the needs of a community are defined by that community or are being defined for them by others. By standing together and truly addressing community needs, the usual purpose of a community food forest will grow into something much more specific to the needs and wishes of the people living where the project is installed. Given this, what is the best way to explore and build opportunities for collaboration and stay true to the general good intentions and useful concepts of a community food forest?

Food forests can be shaped in many ways to address the needs of a local community. If building a community food forest is the most important goal but engagement with the local community is still desired, then bridging social capital must grow. Again, this takes time, but how much is usually dependent on historical context. Some divides run deep and working together on the terms of a community of place is one way to start bridging divides. In order to work toward community buy-in, a willingness to work within limitations is key. Creative planning that captures and responds to opinions, even of those who may not actively join the cause, will be necessary. To that end, curiosity, flexibility, and compassion are attributes that lead to insight and solutions for bridging the gap. Increasingly, places where food forests are located include a mixture of cultures and socioeconomic backgrounds. Only an open mind and balance of pride on the part of all involved will carry the day. And that is not simply a willingness to learn about other cultures, but patience and passion in discovering ways to foster collaboration and bring people together.

The same is true for negotiating the definition and meaning of community between outsiders who are interested in the community food forest and those who actually live there. Conflicts that are triggered by social identity are fueled by emotions about what people believe, how they see themselves, and how they conceive of others. If a group of people that you would like to include perceives themselves as being treated differently than others, it could be a potential fault line that causes larger problems. In contrast, assuming that the minority must adapt and adjust (assimilate) to the majority or the powerful few can equally cause problems. When a local group expresses their distinctiveness and perhaps questions a community food forest, those that are leading the way or investing their time and energy may feel threatened or have mistrust because they do not understand the dissimilarity, even though their intentions are to help.

Thinking critically about potential imbalances can lead to productive brainstorming about ways to proactively defuse them. For example, a community food forest design can include plant species relevant to the cultures of people living nearby. Gatherings can be set up, such as recipe and story swaps, celebrations of holidays of various cultures, or plant walks to explain the cultural significance, uses, and management of different species. Intentionally creating ways for these cultural exchanges to take place in a safe environment can foster understanding and appreciation between different social and economic identities. Along with fun events such as concerts or cookouts, designing joint activities that explore similarities and differences in constructive ways are equally as important as designing species layout. Use the food forest as common ground to bring together people of different backgrounds and from near and far. Working side by side on a project that spans multiple values can result in shared understanding and compassion toward diverse cultures, which is increasingly needed in our growing globalized cities.

WHY COMMUNITY SUPPORT IS NECESSARY

Reflecting on the meaning of community and who belongs to a community creates a clearer picture of who can be approached as potential investors in a community food forest. A network of investors formed through partnerships and collaboration that can harness human, social, political, and financial capital as a means to invest in existing natural capital (land) exhibits richness beyond what monetary returns alone can accomplish. The returns on investment will be much higher. Consider the analogy of buying stock. Think of the community food forest as a stock that goes public; everyone is invited to buy shares using any form of capital they can invest (e.g., social, cultural, etc.). Like a company that performs well in the stock market, a successful community food forest will attract more investors than one that suffers from lack of progress or participation. Dividends will be paid out through percentages to a diversified portfolio of community capitals much like a mutual fund with long-term returns. How much a community invests will determine how much they get out of the project, but what type of capital they invest may transform into returns by increasing other capitals.

Being aware of the flows of different forms of capital between stakeholders helps a community understand what assets they have at the inception of a food forest project, but also where investments are needed, which in turn identifies what stakeholders to work with. A systems view helps visualize how capitals are transforming or accumulating and can lead to a more thorough understanding of the networks supporting the food forest. This is important to consider because most food forests are promoted under the banner that management needs decrease over time while food and other yields increase. It is a compelling notion, but keep in mind that management needs are typically underestimated. Just as it takes years for succession to evolve from grass fields to mature forests, the same is true for a food forest. Planting larger, older trees to begin with can jump-start the process, but a self-maintaining forest ecosystem rarely, if ever, emerges quickly.

For a self-regulating and synchronized system to form, relationships between soil, microfauna, plants, and other components need to be nurtured. Labor and maintenance needs can be time-consuming and are never fully eliminated—they only shift during stages of succession. Food forests mimic a natural ecosystem, but when tied to goals such as food production and environmental benefits, they require additional intervention to ensure reliability. Sites typically need to be managed until they reach a desired production stage, which can take years depending on climate. This is particularly true in urban areas that have been impacted over long periods of time by compaction, garbage, toxins, and alterations such as redirection of water flow. For all these reasons, many community food forests need human intervention throughout their duration to steward a healthy, productive, useful, and safe site.

The limited size of many food forests also inhibits their ability to become self-sustaining. Edible landscaping might be a more accurate descriptor for small food forests, and landscaping requires ongoing maintenance. Once food forests reach the desired production state, they need to be maintained in the preferred condition unless other forms of successional change are a goal. The need for continuous nurturing and maintenance holds true for human communities served by food forests, too, in order to ensure the human community fully benefits. Positive strides within the community, such as volunteer levels and programs, will also change over time, as will the key players. Leaders and planners must accept this reality.

FARGO FOREST GARDEN

The Fargo Forest Garden in Portland, Oregon, is a useful illustration of how to jump-start the transformation of a site into a food forest. Situated on what was once a paved parking lot, this project is now a flourishing oasis. The owner of the lot, Angela Goldsmith, was approached in 2007 by friends who were starting an organization called DePave to help transform paved places to overcome the social and environmental impacts of pavement. They wanted to transform Angela's underutilized parking lot into green space. She agreed as long as the space could be designed so that mowing and maintenance were kept to a minimum. The idea of a community food forest surfaced as a solution, but the conversion would require substantial work. DePave overcame this hurdle by tapping into social networks and other planned sustainability activist events in Portland in 2008. They led a monumental community organizing effort that drew in more than 100 people to break apart and remove 3,000 square feet of asphalt. An intense burst of hard work to transform a paved lot into a planting space is more appealing to many people than the repetitive work of ongoing maintenance associated with keeping an established green space productive.

In tandem with the remarkable group effort to depave the site, Angela experienced a personal

Sweat Equity

For people who are environmentally conscious and concerned about urban justice, investing energy to create change in a short period of time has lasting emotional and psychological rewards. It links people together through social bonding, but once transformation is accomplished, how can momentum be sustained? One way is to evaluate the different roles that stakeholders will play over time to see a community food forest through its phases of development. Those who are motivated to participate in the initial installation may not be as interested in working with it or using it years later. This is demonstrated at the Fargo Forest Garden where the majority of volunteers who helped clear pavement from the site happened to be in the city for reasons other than starting a long-term relationship with the food forest. Team leaders in Philadelphia told us it was much easier to attract a large crew of volunteers for a one-time cleanup of garbage and other deposited waste than it was to get people to show up for routine maintenance. Knowing this and building in strategic measures from the beginning will go a long way in maintaining momentum.

FIGURE 11.2. The Fargo Forest Garden was once a paved parking lot. In 2008, more than 100 people helped remove the pavement, a first step in transforming the site into a community food forest, which is still thriving years later. Throughout Fargo Forest Garden's existence, many groups have been involved in different parts of the process, which means community support has been steady even though there has been turnover among the members of the support community.

transformation inspired by the idea of creating change. She researched food forests for a year prior to planting, studying *Edible Forest Gardens*, by Dave Jacke with Eric Toensmeier, along with other texts and information.[1] This resulted in the design and construction of undulating topography to mimic a forest floor and helped prepare her for explaining the strategy to others. The depaving experience ultimately prompted a new pathway for Angela, which included learning more about permaculture and incorporating it into her work as a real estate owner, as well as becoming a permaculture designer and teacher.

After learning about edible forest ecosystems, she not only designed the food forest structure but also spent time on the layout and composition of a forest floor. By installing undulating mounds across the site that had deeper soil horizons and established an organic top, Angela created moist micropockets for specific plants, provided drier areas on top of the mounds for others, and directed water flow toward areas designed for retention. The payoff is that the food forest now resembles a small forest park with a walking trail through it. Color plate 7 offers a view from above the canopy where the parking lot once existed, as well as a gravel patio hiding a dry well underneath it and the mounded forest floor.

Grant funding for the Fargo Forest Garden specified that the site had to be capable of handling a fifty-year storm event. Creating a forest ecosystem that would thrive on a previously paved parking lot, as well as preparing for major precipitation events, took diligence and innovation during the design process. Extra measures, like the design of the forest floor, helped create moisture microclimates and favorable conditions for plant growth. Holding tanks were purchased with the funding to collect water runoff from the roof gutters of two buildings bordering the lots. When the tanks overflow, they empty into a catch basin surrounded by native vegetation planted as bird habitat. If a rain event causes the catchment pond to overflow, the food forest pathways are designed to direct water to a storm water catchment area hidden under a gravel patio. Below the patio is a nine-foot-deep, six-foot-wide perforated dry well. Water that flows into the dry well slowly disperses back into the site rather than overloading the street sewer system.

The Fargo Forest Garden is also an example of how a group turned a potential obstacle into an opportunity using the versatility of community food forests. Food forests in urban areas must meet additional demands in their design because they often must do more than produce food and provide green space. Funding sources will always have design objectives as their measures of success, and a level of flexibility is necessary to move projects forward. Fortunately, the nature of food forest design and adaptation is one where practical arrangements are combined with creative vision. Working across these interests is one of the best attributes of a community food forest because they are able to strike a balance nimbly between on-site projects and funder interests.

Many stakeholders have passed through the Fargo Forest Garden network in the decade since its founding. They illustrate the complexity of a social system and how it evolves in and around a community food forest through the exchange of capitals. The social system formed when community members concerned about development in the area approached the landowner about converting her parking lot into green space. It is often this initial group that becomes the core strength of the project. These individuals partnered with an organizational stakeholder, DePave, that was also interested in converting the site.

Securing and directing funding from a local government agency toward creating green space within a city is one active expression in shaping the urban landscape. The owners of property that bordered the site also became part of the system. The first floor of a building adjacent to the site is a restaurant and bar. The owners had a stake in community food forest maintenance because they offer outdoor seating to their patrons and harvest herbs there to make specialty drinks.

After the site started producing fruit and other food, two new stakeholders joined to help with maintenance in exchange for access to some of the food supply to support their program, the Urban Farm Collective. This Portland program brings neighbors together to make use of vacant lots for food production, education, and community building. Volunteers log hours when maintaining gardens in a network of gardens throughout the city. They can exchange their labor (hours worked) in a barter system each week at the program's farmers market. The Fargo

Forest Garden is one of the few sources of fruit in the network, and its success opened doors and desire for new community food forests to increase the fruit supply. This is significant because when a project is successful, organizations or individuals have a better chance at winning over critical governmental and nongovernmental supporters for similar projects.

The Urban Farm Collective exists under the umbrella of the Oregon Sustainable Agriculture Land Trust, which donated about half an acre between two major roadways in Portland for the purpose of creating a food forest, now named the Greeley Food Forest. The site is primarily managed to add more fruit options to the Urban Farm Collective's farmers market, but it is also freely open to the public. The food forest is located near an area popular with people in need of shelter, and the collective made a conscious choice to keep the food forest open to those who need it the most.

ORGANIZING PEOPLE

As the Fargo Forest Garden story demonstrates, the way people organize and why plays a major role in community food forest success. Stakeholder identity will change over time, but the need to organize stakeholder efforts remains constant. Typically, a core group organically forms, which lays the groundwork for future successful community food forest coordination. However, the ways in which people understand "organizing" vary, and understanding those differences can make for better strategic decisions and effective social system planning. If you have ever attended a workshop or training on community organizing, you probably realized fairly quickly that community organizing is not the same thing as organizing a community. This section discusses nuances of both to help explain when one may be needed over the other.

Community Organizing

Community organizing is often associated with activism, the action of coordinating and campaigning to bring about political or social change. This type of organizing is typically what drives a social movement. Social movements build a community by organizing groups or individuals with similar values and concerns behind a common cause so they can form a united front and gain power in numbers. In the Fargo Forest Garden example, DePave utilized community organizing to direct the energy of social activists toward extensive initial site preparation. After the pavement was removed, much of the energy stemming from the activists' belief in a mutual cause shifted away from the food forest to other sites in need of depaving. Yet by engaging in this social movement, activists learned about issues related to the cause and increased human capital in the form of knowledge. They learned about how to act in a way that strengthened their presence and shifted the relationships of power. These shifts are connected to social, cultural, and political capital.

DePave is a type of organization that can rapidly mobilize people to take action toward a shared vision of creating change. A defining characteristic is that people are connected through a dedicated network that helps communicate the movement's goals, provides updates on challenges or victories achieved, and keeps everyone informed of opportunities for action, recruiting, and educating others. The art of organizing is directed at campaigning, a strategy of ongoing events, actions, and production of informational resources that brings attention to issues of concern in order to realize a desired result. Leadership roles are determined in relation to getting tasks completed to create the campaign as well as mobilize people. The accumulation of skills and knowledge gained through this involvement empowers people, and though sometimes fleeting, it can profoundly affect the success of a community food forest and its larger social goals.

Organizing a Community

Organizing a community, on the other hand, does not necessarily involve activism or campaigning. For example, one person may organize neighborhood residents for a same-day yard sale so they can share advertising and benefit from increased traffic to the community. With this type of organizing, the community is established by proximity or neighborhood boundaries rather than shared values. A person or a group of people within that bounded space takes a leadership role in organizing others to accomplish an event or project or to change something that directly relates to the community.

In the case of the Beacon Food Forest in Seattle, community organizing was underway in response to redevelopment of an existing city park, a project that took twenty years of campaigning and communication. A group of leaders dedicated themselves to informing the community about rallies, petitions, and more until the change they sought was achieved and formalized. They then built on this momentum and organized community volunteers to help implement the park's new strategy. Later, the leaders of the Beacon Food Forest were able to tap into this network of organized community members and gain extra support for their food forest project. These examples demonstrate that both types of organizing can be involved in the process of establishing a community food forest. It is also important to note that both types of organizing require leadership and collaboration.

INVESTING SKILLS

Thinking in terms of systems and stocks and flows of resources also provides context for understanding how collaboration correlates with coordination of the assets needed to create and maintain a community food forest. The community capitals described in chapter 3 represent assets that are exchanged and accumulated as community food forest currency. Natural capital in the form of vegetation is increased because of human, social, political, and economic investments from a network of civic stockholders formed through partnerships and collaboration. A social system that harnesses these capitals, along with a small amount of natural capital (land) creates richness beyond what financial resources alone can accomplish. Visualizing the relationships between different capitals helps a community understand what assets they have and also where investments are needed. A systems view provides a way of visualizing how capitals are transforming or accumulating and helps build a more thorough understanding of relationships that will support the community food forest.

Collaboration between stakeholders is an example of how diverse investments in community food forests transform and flow through a system. Social capital is one of the most obvious forms of capital that flows between community food forest stakeholders. Investments are substantial when it comes to finding community food forest supporters so that a project can move it into the planning phase. Political capital and financial capital are also important and can be gained through partnerships. The community is built by people of different backgrounds coming together over food, gardening, harvesting, and collective progress. Through all of those situations, how people want to interact and what traditions they seek to honor will emerge.

Power in Systems

In terms of a community food forest, the people involved can affect the system only to the extent that they are aware of how and where to fit in and how they are connected to other elements. Awareness creates a position of intentionality about our role, as leaders and participants in a community food forest, in influencing inputs and outcomes. This consciousness is linked to the principles of a theory of change. Recognizing such a strategy helps us regain power by simply identifying where and how to direct efforts and time instead of struggling and feeling helpless. In *The Dance of We*, organizational development consultant Mark Horowitz notes there are four ways in which a system can be life-affirming and healthy.[2]

- Nourishing people: recognizing that everyone is interconnected and interdependent.
- Honoring life through gratitude, sustainability, intentional relationships, and learning from others.
- Practicing respect for the inherent value of each being and how it expresses life.
- Practicing the double golden rule of treating others, as well as yourself, as you would wish to be treated.

When leaders of the Beacon Food Forest were asked why people participate, they responded that many participants were in transition or going through difficult times and looking for a community that is focused on something positive. They may have been dealing with life changes or looking for direction, and the food

---

Acknowledging and Calming Conflict

Being mindful and willing to question the community food forest paradigm from which you or your group are operating is helpful when confronting challenges. Bringing the values and needs at the root of assumptions and actions to the surface creates a foundation for authentic and open communication in solving challenges and finding common ground. If you recognize that your group is becoming consumed in conflict, a good first step is to acknowledge the conflict. Acknowledging the conflict allows everyone to discuss how it is impacting the group, the project, and overall progress. Collectively decide on a process for handling the conflict—will everyone be involved? Do certain people need to discuss the conflict with a mediator? Lay ground rules for discussing the conflict and exploring solutions so that respect and common courtesies are maintained. Techniques such as active listening or compassionate communication (also called nonviolent communication) are helpful in getting to the root of the problem and making people feel heard. For more information on these techniques and other resources about managing group dynamics, visit www.communityfoodforests.com.

forest offered a landing spot for them. It allowed them to take part in something that is healing and honors life through sustainability, gratitude, intentional relationships, and shared learning. Do you notice echoes of Horowitz's characteristics of life-affirming systems?

To move in the direction of creating a constructive community food forest, we recommend consciously raising and discussing questions about power, because although it is an invisible force, power is strongly felt and affects group outcomes. What is your group's relationship with power, internally and externally? What attitudes do they hold about power? What preconceived notions about power are people bringing to the community, and how might those assumptions impact group dynamics? How will the group balance power when interacting personally and when working with volunteers and other stakeholders? In the case of Richmond, when neighborhood residents asked if they could prune shrubs and trees in the garden, power was shared as a way to add value to their work.

How can support be shown to volunteers that empowers and encourages them to be active in the group? Allowing volunteers to choose how they become involved with a site, whether through physical maintenance, cooking at potlucks, playing music, or attending workshops, encourages participation at levels within comfort zones. The Bloomington Community Orchard (described in chapter 15) intentionally cultivated a culture that honored a person's need to leave the group if it became too much or was not possible given other circumstances. This in turn created a culture of heavily invested volunteer members who wanted to care for the site and organize activities rather than feeling obligated by a mandate.

Failing to address questions of place and purpose at the beginning of a project can lead to undesirable outcomes. At one site we visited, a group acted on their negative views of government authority by attempting to regain power through clandestine activities rather than engaging those authorities in dialogue. As a result of the power struggle, multiple conflicts arose between local authorities involved with the site, the activist group, and others who were also impacted by the behaviors exhibited through the power struggle. Collaboration between groups to maintain and work on the site was strained. Work proceeded at a slow pace, with many obstacles. Opportunities for synergistic partnerships were missed. More than once, detrimental power dynamics caused major disturbance to the site and deepened the divide between stakeholders.

The Richmond Edible Forest Project

Valuing forests and conservation begins with experiences as a child. Such is the thought behind the national Kids in the Woods grant program of the USDA Forest Service, which in 2010 helped create the Richmond Edible Forest Project in Richmond, California. Made possible through multiple partnerships, the project was led by the nonprofit Urban Tilth, which has educated approximately 700 underserved youth by training them how to install edible forests in parks and on school property while learning about the local environment and careers in natural resources.[3] Workshops were provided on edible forest gardening techniques to community gardeners, as well as to school and recreation department staff.

One of the food forests is located in a high pedestrian traffic area in the Richmond

FIGURE 11.3. Urban Tilth's community food forest in Richmond, California, is located in a greenway that receives heavy foot traffic by residents from the surrounding area. Project leaders came up with several successful strategies to involve residents, especially youth. This elderberry has reached canopy size despite the dry climate because of site stewardship by youth who are paid through an experiential learning program.

Greenway, a community revitalization project connecting neighborhoods with a safe public recreational space. The greenway is a rails-to-trails program along the old train tracks and rights-of-way that surrounded it. In 2011, Urban Tilth spent about a year amending and building up about two feet of soil on top of the corridor as a way to avoid toxins in the existing soil. Results of this site preparation are visible in the hearty growth of the plants at the Richmond Edible Forest Project.

Woodland strawberries are used as a ground cover to keep down aggressive weed growth in the rich soil. Elderberry bushes have grown as large as small trees, and residents are often spotted collecting berries to make jam and pruning and tending the bushes. The food forest also includes planting beds for annual crops, culinary herbs, and pollinators. The site has had enormous buy-in from locals, if for nothing more than as a beautiful place where once only weedy growth sprouted from poor soil. Site aesthetics and the opportunity to play and relax in the green space entice people to visit time and again, and inevitably many find themselves checking out the edible plants, too.

The Richmond Edible Forest Project also funded a summer apprenticeship program that hired youth to be stewards of, and participate in, urban agriculture projects, such as caring for the community food forest on the greenway. As part of the program, apprentices get to take part in field trips to places like botanical gardens and go on a camping trip as well. When we visited the greenway food forest in 2015, Sherman Dean, the Richmond Edible Forest Garden Project manager for Urban Tilth, was preparing to take summer apprentices on a national forest camp-

Valuing Volunteers

Leaders at multiple sites stressed the importance of truly valuing volunteers and treating them with regard for their worth rather than simply as sources of free labor. If volunteers are assigned a single task in isolation, their interest will likely wane. In those situations, leaders noted that volunteers will generally feel undervalued, maybe even exploited. A more successful approach is to post or discuss a list of activities that need to be taken care of and let people join in where they feel their strengths and interests are best directed.

Another point leaders emphasized is that the experience should be fun. Creative thinking can combine fun with tasks. At the Richmond Edible Forest Project, kids from nearby homes were invited to help water the site (and play) using water guns to make the task enjoyable and increase youth participation. Allowing children and adults to have fun while getting work done helps validate the valuable free time volunteers provide. It also provides an opportunity for people to strengthen social bonds.

ing trip to help them connect their food forest ecosystem with that of a natural forest. It would also be a vacation of sorts where they could run freely and have fun in nature's playground. Sherman's hope was that as the food forest grows and matures, people who visit will feel as though they have been transported out of Richmond and into a restorative escape. His excitement for the program was contagious.

Sherman said he was thrilled to work for an organization that gave him an opportunity to share with Richmond youth experiences like those he had as a youngster growing up outside the city. He enlisted local youth to help design the food forest. Some of those youth were his cousins, who provided strong ties to the community. They focused on using native species and plants that could withstand drought, so that irrigation could be directed to the many fruit trees during low-water conditions, which enabled events such as the annual multineighborhood pie baking competition to continue.

Urban Tilth's well-rounded program involving the local neighborhood and youth from the larger community is an excellent example of the interconnections and interrelatedness of people, food, and ecosystems. All the examples we describe are testaments to the ways that food forest collaboration can be life-affirming and distribute power and love constructively within a system.

CHAPTER 12

Building Social Systems

Community food forests tend to start at the grass roots. They percolate up from an individual or small group contemplating how to improve their community through environmental projects and food access initiatives. The initial impulse is often to develop a community food forest to provide nutritious local food and to improve the ecological state of underutilized urban areas by turning them into beneficial green spaces. It also usually comes with an

FIGURE 12.1. Chris Rice of Quad City Food Forests explains planting techniques to a crew of volunteers ready to continue installing trees at the Blackhawk Food Forest in Davenport, Iowa. Involving volunteers with site installation is an opportunity to solidify a core group.

expectation that the food forest will become self-regulating. Yet, the sense of community, human relationships, and experience are some of the most important outcomes of community food forests. The idea of giving back to nature and society draws people in.

Community food forests without people are essentially meaningless, because they are participatory landscapes where community and environment become one. Volunteers help shape a natural landscape, produce local food, facilitate community interaction, provide environmental services, improve nutrition, and affect urban agriculture policy. Community food forests bring together people who might not know each other but have similar values and community goals. Through these projects, citizens express their values about public space, civic participation, community life, local food systems, and communal abundance rather than a single individual capitalizing on a resource.

In Philadelphia, the spark that led to a community food forest was the extensive number of unused vacant lots in the city in the face of people waiting for years on a standby list for a community garden plot. In Syracuse, New York, the idea took shape when two individuals discovered that they shared interests in providing health services, fresh produce, and restorative environments to low-income urban residents. One had vacant property next to their free health clinic, while the other was looking for a place to design and install a community food forest.

Once a community food forest idea forms, leaders typically bring together a mix of stakeholders (community members, local government, organizations, religious institutions, and/or universities) to create a shared vision, form partnerships, acquire land, seek resources, and navigate policies. In other words, they build a social system that then establishes the food forest. Building a social system means recognizing that the community support network is at least equally as important to design as the physical site. Ecological design concepts such as redundancy, multifunctionality, and companion planting, along with other polyculture strategies also play a key role in the social system.

Establishing a community food forest is a lot of work but also is typically the most fun and exciting phase in the life of a project. In contrast, planning for a project could take as long as three years, involving deliberate steps that are necessary to ensure the right components are in place before planting. The collaboration that underlies productive planning is catalyzed by visioning together, and it is anchored in communication and partnerships to overcome barriers or obstacles that exist or emerge. For example, such partnerships for food forests located on public property may require formal agreements to safeguard against liabilities and ensure maintenance needs are met over time. On vacant lots, longevity will likely require negotiations with the owner regarding lease agreements and other stipulations.

The concept of resource stocks and flows described in chapter 2 provides context for understanding how collaboration correlates with the coordination of assets needed to create and maintain a community food forest. We have heard from many leaders who felt they had to "figure it out along the way" when it came to gathering groups of people to support the project, finding funding or organizational support, putting governance into place, and engaging volunteers. The aim of this chapter is to provide the tools and strategies to help you avoid that scenario.

BUILDING SOCIAL SYSTEMS

FIGURE 12.2. A group of volunteers at the Mesa Harmony Garden in Santa Barbara, California, meet for a regularly planned work party where they build community. Participants have a potluck lunch, at which time everyone can discuss their week and socialize. They are learning new skills and experimenting with a design for a structure that will "catch" fog for irrigation.

Community building starts by being as inclusive as possible when gathering support for a community food forest, is secured by being strategic when forming partnerships, and is ultimately maintained by being responsible in upholding stakeholder agreements. For example, creating brochures in multiple languages to invite as many people as possible to a meeting is a great first step. Consideration of literacy levels is also important. Do you need to devise another strategy for reaching those individuals? Including visual materials with the text is even more helpful. A second step is making sure translators for different languages are present at the meeting. Forming partnerships with local community groups or neighborhood leaders is another way to encourage participation.

STAKEHOLDER STRATEGIES AND COMMUNITY ZONES

As described in chapter 8, permaculture is useful for structuring food forest elements based on their function in the system and creating conditions that help beneficial relationships emerge between parts of an integrated, efficient, and effective ecological system. Zone analysis originally started with designing the physical features of a property, but as permaculture applications have expanded and evolved, the idea of using zone analysis for social systems, organizations, and landscape or community level design is becoming more common. Toby Hemenway provided many examples in *The Permaculture City* for how zone analysis applies to urban regenerative design.[1] The concept of design according to zones of management is applicable to social systems, too. After all, the zone concept in permaculture is ultimately based on the amount of attention something needs. A similar line of thinking with respect to social systems includes how often you interact with people or organizations, how much energy is required to meet the needs of those relationships, and what is needed to maintain them. Defining different levels of stakeholder interaction and placing each in a particular zone of intensity can make it easier to manage relationships, invest in or exchange community capital, and identify gaps or opportunities for strategic alliance.

Stakeholder analysis is a common first step in developing a project. Some key stakeholder categories are property owners or managers, funders or donors, equipment owners, hired labor or volunteers, regulators, permit or lease coordinators, and site designers. Identifying these key actors early on will help leaders anticipate reactions and develop plans for addressing them. For example, at the Basalt Food Park (described in chapter 10), there were concerns about large wild animals browsing in the community food forest, something most urban areas generally do not have to consider. Project leaders were attuned to this concern, realizing it would be worrisome to important stakeholders such as town officials. In response, they contacted the fish and wildlife department to inquire about what type of fencing and other preventive measures would be prudent and could be budgeted for. This preparatory work helped leaders of the Basalt Food Park address impending roadblocks in their first meeting with town officials, which in turn generated municipal support for the project.

After identifying stakeholders and their potential concerns, it is useful to analyze the amount of power they wield as well as their level of interest and how often they will need to be informed once involved. If you wish to have support from a particular stakeholder, whether a governmental employee or a local resident, critical thinking about what motivates them can help shape and communicate a community food forest concept that resonates and is easier to accept.

A landowner who shares the values of a community food forest is an important early stakeholder to find. The first wave of community food forests were largely established on vacant lots or property owned by individuals sympathetic to the cause. In San Francisco, the 18th and Rhode Island Food Forest was established in 2008 on a vacant lot owned by a medical professional who saw value in the food forest from a public health perspective. In accordance with his beliefs, he offered a symbolic lease to a local permaculture organization. In return, the organization uses the site to teach classes to volunteers and organizes workdays for anyone interested in learning about permaculture techniques and plant selection and propagation.

More recently, community food forests are increasingly being established on publicly owned land, typically under management of parks and recreation or public utility agencies. In such cases, there are two common approaches to take when building an alliance. You can seek out people who work within the agency and are sympathetic to the idea, and then ask them to help guide communication with the rest of the agency. The other approach is to build partnerships and gain support from other nongovernmental stakeholders before approaching a government institution. In Helena, Montana, when resident Jessica Peterson hatched the food forest idea for the 6th Ward Garden Park, she first formed a group of supporters as a way of demonstrating to town officials that multiple people and organizations were interested in partnering on the project.

Regardless of how and why critical community food forest partnerships are formed and whoever or whatever is involved, the need for leaders and other key stakeholders to ensure the most effective use of partner energy and input will be highly important. Working with a group of professionals, volunteers, and community advocates (among many other partner types) requires careful management and specific engagement so as to avoid overburdening them and drawing down social capital too quickly. In that vein, it is useful to revisit permaculture and its increasing role as a frame for designing and effectively nurturing social relationships. Zone mapping concepts can help you think about the social connections and their intensity and timing so that you can strategically plan which stakeholders to involve, when to involve them, and how often interaction will be needed throughout the life span of the food forest.

Zone 00—The Individual

In permaculture, the center point of a landscape is the most inhabited space, the place where the greatest degree of interaction occurs. An individual, a home, or a main building can be considered the center point, depending on whether one is planning a social system or site plan. Why start with the individual when planning a social system? Before looking for community food forest stakeholders or engaging as a stakeholder, it is important to consider what you alone bring to the community and support group. Everyone has some sort of human capital to invest, whether it be a certain skill set, desire to learn, willingness to work hard, knowledge of plants, positive attitude, or simple curiosity. Understanding oneself and managing personal energy reserves, needs, expectations, and interactions are important for maintaining positive dynamics in the core group. We have the most control in managing ourselves rather than external events or other people, and how we conduct ourselves impacts the systems we are a part of and affects others. We can choose to create life-affirming systems for everyone involved. The people initiating community food forests are largely passionate and optimistic and are generally leaders. Their contagious enthusiasm motivates others to act, which is important for the success of a community food forest.

The power of one individual to affect a project cannot be understated. For the Southern Heights Food Forest in Lincoln, Nebraska, that person was Amy Rose Brt. A mother in her midthirties at the time, with a passion for gardening and an ability to inspire others to come together over food, Amy brought a good stock of human capital. She previously had studied landscape design, environmental studies, sustainable community planning, and permaculture in Cuba and worked at an organization called Community Crops, which promoted and supported community gardens and farming in Lincoln as well as other projects addressing local food advocacy and education. One day, she was in conversation with an employee who worked at another nonprofit, Dimensions Educational Research Foundation, which hosts research-based Nature Explore outdoor classrooms for kids. As they were discussing how to make community gardens more active, the seed for the Southern Heights Food Forest was planted. In 2012, the two began to outline ideas for combining an outdoor classroom and a community garden. Amy was interested in community food forests and had a vision for how the two could fit together. Her enthusiasm for the project was contagious, and by virtue of that energy she was able to develop a network of interested community members.

Earlier that year, Amy unexpectedly found out she had a rare form of tissue cancer that was pervasive. Despite the diagnosis, she knew her spirit and passion were things she could still invest in with the time she had. Amy continued working on the design with others and shared her vision with as many as she could before she passed away in 2013.

Her commitment was remarkable and her legacy powerful. She set things in motion and motivated others to carry the project through to completion. Her vision for a joyous place where families come together and share values, children learn about food and explore nature, and people strengthen their sense of community was a lasting gift. Her friends committed to honoring her vision, exhibiting patience and perseverance through a long planning process that included negotiating lease agreements, agreeing upon design elements, and settling many details with the religious institution that owns the property. The motivation to work on behalf of Amy Rose Brt and honor her life spirit provided invisible leadership for the Southern Heights Food Forest and, despite the tragic circumstances, demonstrated the power of a core group that is committed to see a community food forest through to completion.

Zone 0—The Core Group

When researching the project management phases of community food forests, we found that seeking partnerships was consistent during initiation and planning stages. The next step is naturally to start thinking about who else you can get on board. Most leaders begin by casually communicating with friends and acquaintances they think would be interested in providing feedback on the idea. They also want to gauge who would commit time and energy to nourishing the idea and championing it in a way that would make caring for the food forest attractive among the next group of stakeholders.

These preliminary discussions with other like-minded individuals are a first and highly important step in identifying a group of community food forest stakeholders. At every site we visited, we encountered a core group of one to five people (occasionally a few more than five); typically two or three were the driving force behind the project's progress and permanence. They were the "parents" who guided the project through the beginning years. The core group can be compared to tribal elders who protect and demonstrate the values of a community food forest and what it symbolizes. In short, they are the cornerstone of the system.

Zone 0 includes people who interact often and work together to approach stakeholders in other zones as the project matures. The ability to collaborate comfortably helps optimize project initiation and keeps morale high during challenging times. People who are passionate about the project and have faith it can work are excellent additions to the team, as are close acquaintances because they are usually the easiest to convince. A takeaway from our own experience managing a community food forest as well as our visits to other projects is that leaders talk about being strategic in terms of whom they involve in the core group or realize afterward that the reason it functioned well was because of the skills or knowledge of someone who was on the team. Ideally, it is best to know whom you will involve and what role(s) they will play.

Based on our research, we have developed a list, which is not exhaustive, of some of the roles to consider when recruiting Zone 0 stakeholders in light of the unique needs of your group. Along the lines of species selection in a food forest ecosystem, we recommend emphasizing diversity and inclusiveness so the group can serve multiple functions across a variety of skills. Core groups made up of a functionally diverse set of individuals who also share a sense of unified purpose helps ensure redundancy and resilience at the same time. This can pay off substantially in case one person burns out or needs to direct their attention to something else. Below are a few examples of key roles core group members often play.

- Visionary
- Storyteller
- Networker
- Fundraiser
- Champion
- Magician
- Realist
- Action taker
- Motivator
- Party planner
- Time manager
- Researcher
- Facilitator
- Note taker
- Designer

The visionary is usually the person who conceived of the idea, so they are not terribly hard to identify! There will be cases where the visionary has had a strong revelation for the community food forest, but that person may not be talented at communicating to a wide audience. In that circumstance, the storyteller is key. The storyteller will be able to reach some audiences better than others, and a networker might be needed to secure the audience. Once the story has been shared, the fundraiser follows up on leads and has the courage to ask for donations. This person may be a grant writer; if not, the fundraiser needs to work with an experienced grant writer to pursue funding opportunities. The advocate (champion) is a dogged lobbyist for the cause. They have connections or hold a position that can guide project needs through local policies and regulations, package ideas using proper lingo, and speak the language of other stakeholder agencies.

If a group cannot find an advocate, then it should surely look for a magician. We all seem to know someone who is able to get things done with mysterious effectiveness. That person is definitely a good addition to the team. Then there is the realist, the person who understands the idea and supports it and knows how to ground it in practicalities. The realist makes the role of an action taker easier by helping them identify where to get started. The action taker is an organizer, planner, or strategist who can enact the steps needed to help the project unfold, and they typically work closely on logistics. The optimistic motivator is a cheerleader who helps motivate everyone and encourages them to follow the action taker's lead. They can work with the party planner to make sure events are fun and productive and create a convivial sense of community. Despite what you may assume, motivator and party planner are not easy roles. Multiple leaders remarked on the necessity to keep work parties and volunteer activities lively because the balance between work and fun is highly important to keep people coming back.

One person that can help with fun while also accomplishing goals is the time manager. A time manager understands how long a task or event should take, how to gauge a group's energy level, and how to judge when to call it quits or to amplify energy by calling in cheerleaders. The time manager is also crucial during meetings to ensure that decisions are made, tasks assigned, and the interaction flows smoothly. People's time is one of the most important resources they have to invest; in that sense the time manager's role is just as important as that of any financial advisor. During meetings, the time manager works closely with the facilitator or may even take on the facilitator role. The facilitator understands how to form ground rules for meetings and group discussions, keep people on track to produce results, and may occasionally need to be a mediator.

There is also the need for a note taker who may be thought of as a listener, an observer, or a person who is naturally detail-oriented. We think of the note taker as someone who, either mentally or physically, takes notes during a meeting, a work party, a walk-through of the site, or conversations between volunteers. They keep a pulse on what is happening and maintain notes to inform decisions. In addition to a note taker, a project might benefit from a researcher, the person who is willing to put in the time to collect data and analyze results to inform decisions or strategies. Their data can greatly inform grant applications and help recruit new stakeholders.

One of the most important roles is designer. Designers understand the principles and framework of a community food forest layout and know how to use different techniques for gathering and incorporating community input into a final design. We are all designers in some way, and tapping into this mentality will be very helpful when it comes to planning subcommittees, educational programs, and volunteer opportunities, as well as coordinating a plethora of other details in the initial and subsequent phases of a community food forest.

It is instinctual to turn to friends and acquaintances in Zone 0 to gain support during the first round of stakeholder development. Forming a core group is essential to show others that there is interest and to gain traction. Site leaders have also identified the importance of being intentional about who is approached to strategically shape a core group with diverse strengths. At the same time, leaders repeatedly cited inclusivity as critical to the success of a community food forest. One way to strike a balance between these seemingly disparate approaches is to allow people to determine what role they play based on their own perceived strengths. This can also help increase enthusiasm for the project. It is a basic human need to feel that one belongs to a tribe and plays a meaningful role. For those who are unsure about where they fit in but enthusiastically want to join in some way, leaders can help guide them into roles until preferences or strengths for a particular task become more apparent.

Establishing a way to function together as a core group and get things done is a step toward minimizing conflict. Having someone skilled at facilitation is necessary for group meetings and can keep things running smoothly. Though it will undoubtedly arise, conflict does not have to be seen negatively and can help the group grow together if guidelines for handling it are set before it arises. Since the person with the initial idea for the food forest instinctively seeks out supportive friends and acquaintances first to form the core group, working out guidelines early on can help protect friendships, keep the project moving forward, and inform strangers joining the group about the initiative's procedures and culture for working together.

6th Ward Garden Park

Jessica Peterson moved to Helena after finishing a master's degree in social economics in a program with participants from more than eighty countries. The experience introduced her to different worldviews and ways of thinking about community development. She realized that people make decisions based on the resources they perceive they have, and to make smart decisions, people needed to understand what resources they have locally and how to use them sustainably. From this perspective, she believed developing a community food forest where she lived would help build a resilient community, one that perceives access to food more abundantly and would be empowered by knowing how to feed themselves in a sustainable way.

Jessica shared her vision with other people in the community that she knew were working on local food issues through community gardens, food banks, county extensions, and other similar organizations. Her background in community change and organizing helped her strategically build a support team. Jessica put together a one-page flyer describing her proposal and talked to each group to recruit allies. Then she began assembling her core group and picked the first person because of their exuberant energy and excitement about the project. Jessica realized fairly quickly, however, that this person was not the right messenger to deliver the idea to city council. Instead, she made connections with Aimee Teegarden, the director of parks and recreation, who knew how to package and present the concept in a way that the city would understand. Instead of calling it a food forest, she referred to it as a community garden with fruit trees.

Next, Jessica decided to approach Caroline Wallace, an AmeriCorps volunteer at the time who worked with Helena Community Gardens and had a landscape architecture degree. Jessica asked Caroline to get involved because she had unique design skills. She also knew Helena Community Gardens was trusted by the city. So far, in this case, Jessica was the visionary and organizer, and so she looked for a messenger and partnered with a designer. She also approached several organizations that were potential allies.

Jessica was aware of the city dynamics—a small town feel where change was slow and new ideas had to be presented carefully. The people she assembled had the attitude that a new innovative project like a food forest was possible and combined them with organizations that were tried and true and held sway in convincing the city to approve the project. This way, when she was ready to propose the idea to the city, the act of asking would be built on the foundations of organizations with established connections to successful community projects. The approach worked, and the project was easily accepted for the 6th Ward Park without controversy. Eventually, others were purposefully added to the core group due to their time availability, skills in fund-raising, knowledge of plants, and community leadership.

Zone 1—Consistent/Returning Volunteers and Directly Interested Organizations

After the core group forms and a vision for the food forest has been articulated, more support is needed. While the project plan is still being developed, this support might be in the form of asking sympathetic community members to attend meetings, sign letters of support, or help spread the word among their own networks. Getting people interested in the project and recruiting volunteers who will help once the project is underway builds supporters in Zone 1. Connecting with local community-based organizations, nonprofits, or interest groups that have missions or values that overlap with the food forest is often a good way to gain volunteers.

Some of the social movements with local branches, such as Slow Food, Transition Towns, Permaculture Guilds, or Occupy offshoot groups are potential sources for like-minded and motivated volunteers. Master gardeners, native plant associations, foraging groups, gardening clubs, or botanical meet-up groups can also help populate this zone. Local educational institutions from elementary schools to universities may be interested in using the site for educational, recreational, or research opportunities for students. This can be an avenue to a consistent flow of volunteers that can be coordinated by teachers or student groups. In some cases, there may also be a local religious institution or a group of congregation members that are interested in maintaining the site or being involved for social justice reasons.

Social capital plays a large part in forming this network of interested parties that will form a consistent base of Zone 1 volunteers. They provide input on site development, offer time and labor on workdays, become users of the site, and help spread the word to others. During a group meeting after a work party at the Mesa Harmony Garden in Santa Barbara, volunteers noted that the food forest provided a unique opportunity for intergenerational learning and sharing. It was one of the aspects of involvement with the site that kept people coming back. Planning to share meals, listen to music, or simply being together while enjoying refreshments rather than working the entire time helps solidify social connections that builds Zone 1 volunteers. Some of the people or organizations initially in this zone may move into Zone 2 as the project develops, with more limited work at the actual site or direct involvement in associated activities. Recall the example of DePave in Portland, Oregon, and their involvement in establishing the site. After the initial workday, the group transitioned to a different zone where participation was limited to using the site as a demonstration.

People in Zone 1 typically stay motivated and continue participating at a consistent level, but understanding what is cogent at this level can shape programming that naturally keeps them involved. For example, Master Gardeners must fulfill a program requirement to serve volunteer hours and help educate the public, but forest gardening may be a new concept to them. Some investment in preparatory training for the benefit of both Zone 0 and Zone 1 groups will enhance the exchange of information and education.

Zone 2—Occasional Volunteers and Indirectly Interested Organizations

Zone 2 will likely include a similar mix of people as Zone 1, but how they are engaged and involved fluctuates relative to how often they are available or their level of interest. Organizations such as food banks that benefit from work being done at the site but do not directly have members involved are an example of a Zone 2 participant. There may, however, be occasional or seasonal events planned cooperatively with these groups. For example, in Troy, New York, the Sanctuary for Independent Media partners with organizations such as Troy Bike Repair on loosely collaborative projects that train youth on bicycle repair and building. The program entitles participants to a bicycle after they have helped train others about what they learned. As a show of support for the organization, the Sanctuary incorporated gravel bike paths into their networked space of gardens and food forests so the youth have a place to ride. In return, Troy Bike Repair supports the Sanctuary by encouraging youth to use the trails and get involved with the Sanctuary's youth programs, which include learning how to grow food.

Zone 3—Peripheral Agencies

Agencies, institutions, or organizations that support the project through funding, liability insurance, leases, or a memorandum of understanding are generally in Zone 3. The amount of time and energy the core group will need to expend with this zone is simply enough to keep a relationship in place. For funding agencies, that translates to substantial time and energy up front to prepare and receive a grant, but during the use of the grant, involvement will probably be limited to occasional updates and periodic reports. There will also most likely be a requirement for a final report that explains how funds were used to meet original objectives.

Quarterly, semiannual, or annual meetings or reports to stakeholders in this zone are the extent of demand on attention. In some cases, an organization that offers liability insurance to volunteers might also offer workshops or trainings that can move a stakeholder into Zone 2. Such was the case at the Beacon Hill Food Forest in Seattle, where relationships across zones strengthened over time, and some key individuals and organizations increased their involvement.

Another example is the Hale-Y Community Garden Food Forest we manage in Blacksburg, Virginia. A funding department within the local university provided a grant to initiate the food forest. Funds needed to be used within a year with quarterly reporting, a final report, and one community presentation on the project. A year passed after the final report where no involvement from the funding department occurred. In the third year, the funding department supplied another small grant to continue development with the same reporting requirements. The funding department also helped connect us, as site leaders, to volunteer groups willing to help with maintenance. From that pool, some volunteers became Zone 2 participants. A useful strategy is to be always on the lookout for how organizations or events in Zones 3 and 4 can be used as catalysts for producing people who will join, and hopefully stay, as part of Zones 1 or 2.

Zone 4—One-Time Events or Engagements

Zone 4 is mostly composed of public relations and media; it usually requires the least amount of attention. Core group members might have to engage in interviews or lead a site tour to provide basic information for an article. The amount of energy generated from such efforts is usually minimal, but every once in a while, it can be dramatic. If a media piece is well executed

PURPOSEFUL COMMUNITY

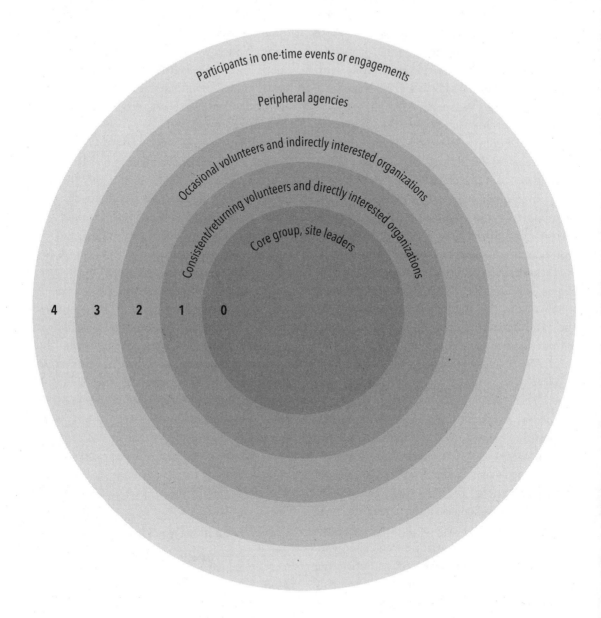

FIGURE 12.3. This zone template offers a basic way to map the social system of a community forest. It can be used in planning workshops, events, and maintenance activities. Zone 0 represents the core group, which is typically small. It includes the people who help manage the site, volunteer activities, budgeting, and other basic operations. Zone 1 is the largest: Having as many consistent volunteers as possible is important. For more information on how to use this template to plan the social system of a community food forest, visit www.communityfoodforests.com.

and the right people read or view it, substantial attention can be the result, as was the case with the Beacon Food Forest, in Seattle, which was not prepared for all the media attention they received and found themselves spending more time addressing media and less time leading progress at the food forest. Zone 4 activities may result in something potentially overwhelming! In fact, the Beacon Food Forest group had not sought an interview or looked for someone to write an article, but the media spiral launched nonetheless. Influence from this zone can surface from many directions, though it may not require sustained energy. Nevertheless, it is important to think about how to direct it properly, such as into a fund-raising event or community festival to recruit new volunteers.

We struggled with the idea of a fifth zone associated with virtual connections, but ultimately concluded site leaders need to best decide where to place these supporters. The influence of people and organizations linked to a site through social media or virtual communities cannot be underestimated. We are all well aware of the power of an online presence and how it helps organize people around an idea. Virtual interest in a site can be creatively leveraged and transformed into palpable support; you will find an example of this in chapter 15 when reading about the Bloomington Community Orchard. Site leaders tapped into an extended virtual network connected to the town and local university to gain online support that helped them win a grant that supported installation of the food forest.

CHAPTER 13

The Beacon Food Forest

One of the most visible and well-known community food forests in the United States holds that distinction not necessarily by choice. A media barrage in the midst of planning vaulted the Beacon Food Forest onto the international stage, leading to extensive publicity on the project itself and food forest concept more generally. Somewhat unexpectedly, leaders found themselves beset by numerous requests for interviews and site tours, which posed a serious challenge because no plants were even in the ground yet! At the same time, however, the Beacon Food Forest demonstrated that due diligence and perseverance before fame is crucial for weathering exposure and managing the unforeseen.

The Beacon Food Forest revealed that, in the midst of the unexpected, projects built from the ground up using strategic community capital investments possess the moxie and people power needed to survive despite diversion. Beacon team leaders focused on building public and private community support over several years, which buttressed the project as it evolved in both incremental and punctuated ways. They had monetary support from the city of Seattle and benefited from a solid volunteer base. The core group shared strong bonds, and the level of project engagement cut across multiple stakeholder zones and involvement. A volunteer

> **Fast Facts**
>
> OWNER: **Seattle Public Utilities**
> LOCATION: **15 Avenue South and S. Dakota Street, Seattle, WA 98108**
> PLANNING STARTED: **2009**
> ESTABLISHED: **Site preparation began in 2012. Planting of Phase I took place through fall 2013 and 2014. Planting of Phase II is scheduled to start in summer 2018.**
> SIZE: **3.5 acres total; 1.75 acres planted as of January 2018**

training program helped build participation and was also a ladder for growth and development toward more serious roles in project leadership. Such strategic thinking about how to distribute power in the system and create leaders represents a highly efficient and meaningful denizen participant model. They have even had highly effective relationships that were one-time events. As news about the Beacon Food Forest traveled around the world, an organization based in Taiwan invited the Beacon leaders in 2014 to their country to give workshops on how to establish a community food forest. The first community food forest in Hsinchu, Taiwan, opened in 2016, proving that making and keeping connections in the outer zones of a social system can be equally as impactful as maintaining those that require more consistent time and energy.

ABOUT SEATTLE

Before settlers arrived in the Northwest region of the United States, the lush, dense forests of the Northwest coast supported at least thirty tribes of Native Americans. The state of Washington is known as the Evergreen State for its thriving, year-round greenery in the temperate forests of the western part of the state. With magnificent growing conditions, it is no surprise that one of the state's leading industries is timber, along with other perennial species such as apples, pears, grapes, and cherries and a thriving nursery businesses. The city of Seattle, located along Puget Sound, has a Mediterranean climate and was given the nickname Emerald City in the 1980s because of its verdant landscape and extensive parkland. The city is within USDA plant hardiness zone 8b, which means the average annual minimum temperatures can be between 59°F and 68°F (15°C to 20°C). Microclimates exist throughout the city due to pockets of warmth from pavement or buildings, wind off the water, and elevation. The result is warm, mostly dry summers and cool, wet winters with little snowfall.

Seattle was originally called New York-Alki by the settlers who founded it in 1851. *Alki* is a Chinook word meaning "by and by," and the name of the city stood for the hope that by and by Seattle would become the New York City of the West. Seattle had an advantage over New York, though. Because Seattle was founded later, city planners were able to learn from developmental mistakes in the Eastern cities. When Seattle was planned, designers pointedly incorporated abundant parklands before a growing population could gobble them up for private or commercial uses.

Seattle is the twentieth largest city in the United States, with a population at around 668,400 people, but it ranks sixth in terms of population density per square mile. The favorable growing conditions combined with high population density create conditions for a thriving plant nursery industry and green community projects such as community food forests. One such project is the Beacon Food Forest, which as described earlier sparked major community food forest publicity, inspiring other towns and cities across the country to follow suit. Within Washington, it catalyzed replication in Spokane (the Inland NW Food Forest Council Partners in Spokane), and spurred old orchard reclamation in other parts of Seattle (the Puget Ridge Edible Park).

IDEA GENERATION

The Beacon Food Forest was founded by Glenn Herlihy and Jacqueline Cramer. Glenn was the site historian, and during our conversations, he recalled the struggles, challenges, victories, and eventual transformation of the site and organization. For more than twenty years, he was an estate gardener in Seattle before starting his own business designing and installing gardens for clients. He was also once a community project activist and an artist who weaved nature into the pieces he displayed around the city.

Given Glenn's interests and experience, a friend suggested that he might enjoy a perma-

The Olmsted Legacy

The vision for Jefferson Park was developed by the sons of Frederick Law Olmsted in an era when it was mandated that every resident of Seattle be located less than half a mile from green space.[1] In the garden city aesthetic, Jefferson Park was intended to be an escape from the busy downtown and an oasis of tranquility with stunning views. The land was purchased by the city in 1898 and has existed as a public park since the early twentieth century, although over time it has been reduced in size. The Olmsted family had a hand in developing many city parks throughout the country. Archived documents with the original visions and uses for the parks may provide a historical rooting for the incorporation of a modern community food forest to recognize or even restore elements a historical vision.

culture design course. Glenn took the course, which helped stitch together his knowledge of perennial plant landscaping with ecological design, landscape creativity, and food-literate culture. While he was familiar with the concepts that shape a forest garden, it was only during the course that Glenn first heard the term *forest garden* or *food forest*. The main instructor for the course was Jenny Pell, an enthusiastic, articulate, and passionate woman who sought to create local places of abundance for future generations. She was helping to lead Permaculture Now!, an organization that designs projects the world over.

Jacqueline Cramer attended the course with Glenn. With her background in farming, education, planning, landscaping, and community organizing, Jacqueline was a natural fit as cofounder of the Beacon Hill project with Glenn. She had operated and worked to certify three organic farms and was a nutrition educator who helped maintain food gardens at numerous schools. Jacqueline and Glenn, along with two other students, worked together on the course's final design project, which set the wheels in motion for the Beacon Food Forest. Their project focused on a piece of land bordering Jefferson Park in Glenn's neighborhood, Beacon Hill, and thus the name.

Shortly after moving to the Beacon Hill neighborhood in the mid-1990s, Glenn joined the Jefferson Park Alliance, which focused on Jefferson Park, the sixth largest park in Seattle and a historically significant landmark. At the time Glenn became involved, two unused reservoirs surrounded by barbed wire dominated the western portion of the park and sullied the view from the overlook for which the park was known. One of the reservoirs was scheduled to be filled; the other would be rebuilt. The Jefferson Park Alliance advocated for filling the second reservoir with a hard cover, so the area could be used for recreation and other community activities. To help make their case, the alliance coordinated an extensive community effort to reclaim the park as a useful and appealing green resource for the Beacon Hill community. Through their advocacy, the group helped secure $8 million through the Pro-Parks Levy to have the hard cover installed on the second reservoir and recreational opportunities created above it and in other areas of Jefferson Park.

In the early 2000s, the portion of the park containing the reservoirs was given to Seattle Public Utilities (SPU) to manage as water conservation land. Through Glenn's work with the alliance, he became aware that an arboretum had been planned for the southwest slope of the reservoir, but was eventually dropped in favor of a mowed grass landscape. In mid-2009, the Seattle city council was discussing a resolution to prioritize protection of the city's tree canopy along with legislation to create an urban forestry commission to advise the mayor and council on related issues. Glenn saw an opportunity to turn the scrapped site into an edible arboretum that would serve the local community and increase urban tree cover.

The group of four permaculture course students focused their final project on designing a site that included walkways among slopes covered with trees and multiple small holding ponds for water runoff, along with a gazebo and pavilion. The design was inspirational and though many hurdles existed in actuality, such as water conservation restrictions that prohibited standing water at the site, their effort was the beginning of work that would lead them and others on a remarkable journey.

BUILDING SUPPORT AND FORMING PARTNERSHIPS

After finishing the course, Jacqueline and Glenn discussed whether they wanted to move forward with food production at the site. They were aware that the community desired some sort of garden, but other than their course project, no formal ideas existed, let alone designs or funding. After much discussion, they decided to host a community meeting to present their idea and gauge support. Using all the email lists and connections they could round up, the team invited as many people as possible to a meeting on Groundhog Day in February 2010 near the property in Jefferson Park. They presented their design, then known as the Jefferson Park Food Forest, and shared ideas about how it aligned with the North Beacon Hill Neighborhood Plan adopted by the city in 2000. The plan called for terraces covered with trees and other natural vegetation, pedestrian pathways, accessible views, and educational facilities.

Around thirty people from the community showed up to the meeting, along with members of local organizations, the parks department, and a few community groups interested in food and the food forest idea. The response was overwhelmingly positive, and the experience inspired Glenn and Jacqueline to move forward. Attendees formed the first interest email listserv (remarkably, some still serve as steering committee members today). That was the first step in building a core group. In late winter 2010, Glenn shared the community's positive response on the Beacon Hill blog, along with an explanation of the food forest proposition. There were additional opportunities scheduled to help residents learn more about the plan and provide input, and momentum started to grow.

In the summer, a presentation about the food forest during an annual festival in Beacon Hill, the Jefferson Park Jubilee, introduced the concept to even more community members. Even though Glenn was not fully comfortable holding a public position or leading community meetings, his belief in the importance of the project motivated him again and again to push it forward across various community contexts. Persistent community outreach was the key to effective management of Zone 2 in their social system, with some of those people eventually becoming some of the most active Zone 1 stakeholders.

Seattle city council declared 2010 the "Year of Urban Agriculture" as a way to celebrate and promote five years of effort to increase commu-

THE BEACON FOOD FOREST

FIGURE 13.1. A simple lunch of healthy food provided by Beacon Food Forest at the end of a work-party day. Potluck items are contributed by community members as they are able. Photo courtesy of Jonathan H. Lee.

nity access to locally grown food. Earlier, in 2008, the council had adopted a local food action initiative that aimed to improve the local food system and build a more sustainable and secure regional food system. However, action beyond adopting the initiative had not occurred by 2010. This led to the announcement that 2010 would be the year to enact initiatives such as developing a new community garden through the city of Seattle's parks levy grant with new land use codes that would support urban agriculture.

The timing was right and the Beacon Food Forest steering committee used this to their advantage by promoting their plan as direct action in accordance with the 2008 Local Food Action Initiative. The Year of Urban Agriculture seemed like a promising time to start the food forest, but in reality, Glenn, Jacqueline, and a handful of core members known as Friends of the Beacon Food Forest would have to endure three years of bureaucracy and negotiations with the city before gaining permission to establish the food forest. They would spend most of 2010 building the foundation for community and political support. The next year saw their focus shift to site preparations through soil amendments and mulching in anticipation of planting a forest and establishing garden plots.

Dialogue began around the same time with SPU, who owns the land, and the parks department, which maintains Jefferson Park. The two agencies and the Jefferson Park Alliance met on site to discuss the food forest plan in detail. The conversation was positive, but no formal commitments were made. There were concerns about restrictions on water conservation land, the need for an official design by a certified Washington State landscape architect, implications of long-term maintenance, and the magnitude of community support required for the project. The public utility was not set up to handle the specifics of such a venture and had no experience with agriculture-related projects on their land. Codes and regulations had to be rewritten to create policies for the planting of edible plants managed by a community group. Facilitating the effort of one community group to install a project on utility property prompted concern that the utility would be seen as giving preferential treatment to a select group. As a public utility, SPU needed to remain responsive to multiple stakeholders.

As other food forests leaders have noted, it is highly important to have a champion for your cause within the local agencies—in other words, someone who will advocate for changes until they are achieved. Having an ally on the inside also helps in crafting descriptions of project needs using language that will be recognizable within an agency or other allied organization. The Beacon Food Forest won allies over time, but in the beginning they had to navigate agency language, decipher regulations, and craft project descriptions by doing their own research and learning what works best.

The Legendary P-Patch Program

SPU required that Beacon Food Forest, which was an all-volunteer project supported and managed solely by the community, find an umbrella organization that would help assure project maintenance over time. Several organizations were vetted, and the Seattle department of neighborhoods P-Patch program was deemed a good fit for multiple reasons. The P-Patch program has existed for more than forty years and was the largest municipal community gardening program in the country, including nearly ninety P-Patch gardens throughout the city on about fifteen acres of land. P-Patch gardeners also maintained fifty-two "giving gardens," which are plots dedicated to growing food for donation. Individual gardeners can choose to contribute a percentage from their personal plots as well.

The organization's grassroots origins and desire to give back to the community were characteristics that aligned with Beacon Food Forest's values. By incorporating as an official P-Patch, the Beacon Food Forest gained an affiliation with a community-focused program underpinned by a long, successful history. Membership dues for the privately owned plots in the P-Patch section of the Beacon Food Forest would help cover recurring fees such as water usage, and community gardening guidelines were already in place. The food forest would also have access to GROW Northwest, a nonprofit that focuses on building healthy and diverse communities that support urban agriculture and green spaces; it was formerly called the P-Patch Trust. GROW would provide liability insurance for volunteers and act as the fiscal sponsor for future grants.

With partnerships in place and the city's 2010 focus on expanding urban agriculture underway, the project had access to major grant opportunities through the Seattle department of neighborhoods to begin installing the food forest. Throughout 2010, the Beacon Food Forest steering committee wrote multiple

Partnering with Established Organizations

Finding established organizations that are willing to back a project can be a great first step in jump-starting agency support and a useful way to tap into existing networks and resources. In the case of the Beacon Food Forest, GROW had a long history as an advocacy and support organization for the P-Patch program. Partnering with GROW gave the P-Patch program access to multiple resources and connections. GROW works with the city of Seattle, but because it is a volunteer-driven organization, it is able to use more than 80 percent of donated funds directly on projects and a wide range of services. Those services include, but are not limited to, acting as a land conservancy that can own property to ensure gardens in perpetuity, developing site leaders, buying tools and seeds for gardens, providing liability insurance, providing small garden development grants, helping low-income gardeners with fees, and facilitating garden donations to local food banks. GROW also provides fiscal sponsorship, a legal mechanism the food forest needed to raise and manage their own funds. By forming partnerships with the P-Patch program and GROW, the Beacon Food Forest tied into a system that was primed to establish and support their initiative by providing a pathway to funding, the credibility of an established program, and knowledge resources.

grant proposals including an application for a $100,000 Parks and Green Spaces Levy grant. Steering committee members attended neighborhood events and set up tables at local grocery stores to build a mailing list of more than 400 individuals. This tangible support led to a small grant in December 2010 that helped begin installation in earnest.

The award was for $22,000 in the "Small and Simple" program associated with the Seattle Department of Neighborhoods' Neighborhood Matching Fund. Funds were to be matched one-to-one by the applicant through services, cash, or volunteer time. The purpose of the grant was to create outreach materials for gathering community input and support, along with hiring a professional team to guide the design process as required by the P-Patch program. The steering committee considered multiple teams to work with, but the Harrison Design team won the project bid in early 2011. The team was composed of several professionals and business people, along with Jenny Pell.

Community input and support would take active and purposive engagement in highly diverse neighborhoods. The census tract surrounding Beacon Food Forest is divided into five main ethnic populations, and approximately 12 percent of the population lives below the poverty level. To build lines of communication between the community members and project leaders, the steering committee printed 6,000 postcards in five languages and mailed them to zip codes surrounding the site. They also created posters and scheduled information tables at local events, grocery stores, and other venues.

Soil Is Life

The P-Patch program is well aware of how important soil health and site preparation are for garden success. The Beacon Food Forest site, however, was in disrepair following earthworks at the reservoir. Most of the topsoil had been washed to the bottom portion of the site during rainfall. The compacted soil that remained created undesirable conditions for plant life. Hiring the Conservation Corps provided a quick way to overhaul the site without fatiguing volunteers who would be needed later for planting and other support.

Preparing a food forest for planting can be hard work depending on its previous use. Looking for partnerships such as the Conservation Corps or other similar paid internships is a good way to bring in a group specifically set on site preparation. (Recall the example of the Fargo Forest Garden, where the DePave group supplied a mass effort of labor to remove paving material.) Earmarking a team or major event dedicated to site preparation can quicken an otherwise long and strenuous task.

Rehabilitating soil is not a one-time task or something easily accomplished overnight, so resources and volunteer groups are generally needed on multiple occasions. In 2014, the Beacon Food Forest arranged for forty-five loads of compost, mushroom inoculant, and compost tea to be delivered to the site along with eighty-plus yards of wood chips. It required hard work from volunteers to incorporate the amendments, but it also provided a great learning experience for volunteers on how soil and forest floor are critical components of a community food forest ecosystem. The results are evident in the rapid plant growth and knowledge among site volunteers.

To begin a garden in the P-Patch program, one must host three public meetings to gather community input. The design team helped facilitate these meetings to collect ideas for layout and plant choices. Large sheets of paper depicting the site design were placed on tables, and participants were encouraged to write in their ideas, concerns, and comments. Translators at the meetings facilitated participation in Spanish, Somali, Chinese, and Vietnamese, which were predominant languages in the neighborhood.

The meetings successfully showed community support, with 70 people attending the first, 95 at the second, and more than 120 at the third. A year after receiving their initial "Small and Simple" grant, the Beacon Food Forest received $100,000 from the Parks and Green Spaces Levy grant program to pay for the final construction blueprints, permits, water aim, and irrigation system. The grant also allowed the group to hire the Conservation Corps to assist with much needed regrading and hardscaping of the eroded and compacted soil left behind following heavy construction needed to fill the reservoir.

Having areas that were hardscaped and leveled was important for installing walkways throughout the site, individual P-Patch garden beds, and meeting and food preparation spaces. Because the property sits on water conservation land, no permanent structures are allowed, which posed a challenge for the proposed sheltered space that would house teaching and community gatherings. The solution was to lay a thick surface of permeable stone and erect wooden structures without foundations. Restrooms and trash cans were also not allowable, nor was the use of fresh manure for organic fertilizer or holding ponds like those proposed in the original design. The list of things that were not permitted was long, but the Friends of the Beacon Food Forest found creative ways to move forward.

INSTALLATION AND COMMUNITY ENGAGEMENT

During site preparation in 2011, the Friends of the Beacon Food Forest were still mostly mired in bureaucracy and working hard to gain permission to plant on the site. Even members of the Friends were starting to suggest other sites for which the permissions process might be simpler. Glenn remained steadfast that their work was focused on the right location and that urban agriculture, open to the community for harvesting, needed to be seen in this location as an example of what is possible on public land.

In early 2012, as final negotiations were taking place on what would be allowed on the site and permissions still being acquired, the media caught wind of the project and publicity about it spread like wildfire. Crosscut, an online independent nonprofit electronic journal for the Pacific Northwest, was the first to write an article describing the project, its potential, and the challenges it had faced. *Seattle Weekly* picked it up the same day. Within a month, national and international news sources were covering the story, including the Associated Press, NPR, *The Salt*, Grist, Mother Nature Network, Gawker, and many others. The core team started receiving phone calls and emails from around the world requesting information, interviews, photos, and public materials about the "first of its kind and largest" public food forest!

Despite the sudden onset of publicity, what was not being reported, and what most of the public did not know, was that the project was *not* the first of its kind. Asheville's Dr. George Washington Carver Edible Park was established in 1997. San Francisco and Portland, Oregon, started sites (although on private property) in 2008. Bloomington, Indiana, created a community orchard on public property in 2009, which evolved into a food forest over time. And although three years of planning had already transpired, Beacon Food Forest remained only a concept on paper rather than a functioning site with plants in the ground.

Glenn tried to answer as many inquiries as possible, but all he could do was explain the idea and share written plans. The massive response to the idea seemed to indicate positive public attitudes about using urban public space for communal perennial food production. It also demonstrated the power of the internet as a means for spreading ideas quickly and globally, which stood in stark contrast to the time and place in Asheville when the Dr. George Washington Carver Edible Park was established in the 1990s.

The swell of attention increased the Friends of Beacon Food Forest's resolve, and as the stories of their food forest filtered through the media, important public dialogue ensued among people responding and reacting to online articles and news outlets. The feedback illuminated ideas that needed further thought. For instance, the P-Patch program, along with other community

Controversial Commons

People around the world were enthralled with the communal idea of the Beacon Food Forest, but many questions emerged about its ability to succeed. The group had to weather an onslaught of criticism from those who believed the food forest was bound to fizzle, based on concepts such as the "tragedy of the commons." Similar questions have been raised in relation to almost every community food forest we studied. If you are involved with a site, be prepared to discuss pros and cons of the commons at some point.

The tragedy of the commons was made famous in an essay by ecologist Garrett Hardin in 1968.[2] Hardin suggested that some participants in communal endeavors act in their own self-interest even if it is contrary to and at the expense of the common good. The combined impact results in the depletion of community resources and ultimately the failure of the community as a whole. But Nobel laureate Elinor Ostrom argued that the tragedy was not inevitable. Ostrom shared eight principles for creating a sustainable and equitable system of commons governance. They include things such as basing rules on local needs and conditions and ensuring those affected by the rules can modify them and create a system for monitoring behavior carried out by all community members. Perhaps the most important and likely most difficult to enact in a community with a mix of temporary and long-term residents is "building responsibility for governing the common resource in nested tiers from the lowest level up to the entire interconnected system."[3]

Looking at other projects that manage commons can help provide an understanding some of the potential solutions to challenges that others have leveraged before encountering the same issues in your project. See www.communityfoodforests.com for more resources on the commons.

garden programs, offers one plot per family or individual, which is a different model than growing food collectively as a community in an area open to everyone, regardless of whether they helped tend, harvest, or manage the site. Through this public dialogue, Beacon Hill stakeholders came to see a real need and desire within their own group to lead a truly public, community food resource rather than a set of community garden plots distributed to lucky recipients.

The Friends of the Beacon Food Forest realized they would need to take a look at policies that could potentially affect a public food park rather than a community garden. They knew that in order to enact communal management, the public would need a sense of ownership of the project.

The media attention resulted in a huge increase in volunteers from the Seattle area who reached out to inquire about getting involved. In the summer of 2012, the Friends of the Beacon Food Forest started advertising that they needed people to join and learn about leading planting and work parties scheduled for later in the summer. At the 2012 Jefferson Park Jubilee—two years after the initial presentation about the project—the committee placed markers and flags at the site to help people visualize where features would be located and to cultivate community interest in anticipation of breaking ground at the end of summer.

That September, more than 100 volunteers showed up to help install the first trees, bushes, and growing beds. In addition to the volunteers, multiple community organizations helped transfer plants to the site, feed volunteers, and attend to other logistics. The following day, there was a workshop to train a small number of neighbors in planting and caring for more trees. Throughout the fall, additional workdays focused on digging irrigation trenches, sheet mulching, and additional site development. P-Patches and fruit trees were installed first, understory plants were installed later, and annual plants were given room to grow in open areas within remaining space.

The seven-acre site was split into two to stagger planting, and Phase 1 was implemented in segments and layers. A dessert potluck in early 2013 was kicked off by a celebration of the previous year's accomplishments and a viewing of the movie *The Power of Community* to get people excited about the coming year. The movie chronicles how Cubans created agricultural solutions to feed themselves despite challenging politics.

The event was also a chance to celebrate a $12,600 grant from the nonprofit organization Sustainable Seattle to create educational signs. Meetings were held with the community and two designers to collect input on what types of ecological and community information should be displayed. The signs were installed that July in multiple languages based on the major ethnicities represented in the neighborhood.

In March, open public meetings were held to collect ideas and opinions on the design of a gathering plaza. In April, another big news article came out titled "Seattle's Free Food Experiment" in *National Geographic*'s online magazine. It sparked yet another wave of publicity, which was picked up by multiple online sources such as Modern Farmer, Mental Floss, and Nature World News to name a few. By the end of the year, more than thirty-five organizations, associations, academic institutions, and local companies had partnered in some way with the Beacon Food Forest committee.

By 2013, the committee had divided into steering, site development, and community engagement subcommittees to handle the growing number of events, grants, and site-specific tasks. Public interest remained high through 2014, which saw more than 2,000 volunteers participating at the site to accomplish Phase I of the installation strategy and finish funding from the Department of Neighborhoods Large Matching Fund grant, which meant all fruit trees, berry bushes, and perennial ground covers had to be planted. Other installations included a mushroom hut, a composting area, an herb spiral, pollinator patches, a gathering plaza, a kitchen building, an emergency hub, a trellis, and benches.

Another installation was the Common Thread garden area, which was planted to honor the ethic of fair share within the community. A wide variety of vegetables and herbs were planted, and more than 445 pounds of food were harvested to share with local neighbors, along with an additional 407 pounds from the Common Thread garden, all of which was combined with other communal area harvests and donated to the food bank at El Centro de la Raza. As Glenn noted in the 2014 annual report, "Sharing food comes with great rewards as it builds community integrity, trust, and resiliency and demonstrates equality and food

FIGURE 13.2. The gathering plaza at the Beacon Food Forest serves multiple purposes. In addition to being a place for potluck meals to feed volunteers during work parties, the plaza harvests rainwater from the rooftop using barrels and the door is covered in chalk paint to serve as a message board.

justice by providing equal opportunity for public land stewardship. Building community is more important: How much yield we produce is almost as important as how much community we build and educate."

Community building, education, and land stewardship have definitely been major accomplishments realized by the Beacon Food Forest. Various workshops are provided on site and learning opportunities are woven into every work party. Maintenance at Beacon Food Forest is always steeped in education. They make sure that workdays are work parties, and throughout the year, they hold permaculture training collectives on the site. In 2014, volunteers bolstered the soil on the site and developed a watering system to keep plants alive through the summer using as little water as possible by continuously training volunteers on how to check soil moisture levels. With education as a consistent focus, more than fifteen tours of the site were given to students and sustainable agriculture groups. Research on pollinator diversity by local students has contributed to a richer understanding of how the site relates to nonhuman life and mimics a natural ecosystem.

Volunteers are eligible to submit a proposal for hosting a training, as are nonvolunteers who have experience with permaculture. The Beacon Food Forest core group evaluates communication skills and demonstrated knowledge of the applicant. Guidelines are provided on expectations for the workshop structure so that there is consistency among the events offered. Proposed topics range from collaborative social change, medicinal plants of the food forest, and basic herbalism to graywater systems, seed saving, starting a garden, and fruit tree pruning. A stewardship program is taught quarterly by local permaculture educators; it consists of three classes that address ecological theories and applied methods of food forestry coupled with time for discussion with site leaders. Participants are required to take on a leadership position at work parties to put their new knowledge to use. Free harvest dinners are hosted at the end of every season to enable the public to partake in the bounty of the larger community collaborative.

Even with all the care and attention given to soil, plants, education, and community outreach, the project has of course faced challenges and had to work through change. Some of the plants have died or struggled and tree labels have gone missing. By 2016, a few trees were beginning to show signs of overharvesting during periods of early growth. The committee sees these challenges as teaching opportunities and places for improvement rather than reasons to quit.

CONTINUATION AND FUTURE VISIONS

The leaders fostered a sense of belonging by engaging volunteers on a regular basis and empowering them during each workday to take ownership of small projects that contribute to larger installation aims. People felt invested in the food forest's success because they contributed their time, effort, and emotion into building a shared vision. Glenn and others were amazed at the diversity and sheer number of people that regularly showed up to help. Community outreach included providing information in a variety of languages to attract people from the surrounding neighborhoods. However, people interested in the concept of a communal food forest, with a range of motivations from political to environmental to personal, were the first to respond.

While the goal is to include as many community members as possible, those in the core group understand that the neighborhood has people who might be able to show up only some of the time. By providing numerous ways to get involved with the site, including multiple celebrations throughout the year, they hope neighbors will find a way to connect and create friendships. It often takes a community of people interested in the project to establish a site, but for long-term sustainability and goals of social justice, methods of engagement and governance involving people from the local neighborhoods are critical. In fall 2016, the food forest took steps forward in this regard by bringing in speakers for workshops on how to create cross-cultural dialogue within diverse communities.

Their emphasis on community building as the most important yield of the food forest helped them secure a grant in September 2016 from the large projects fund of the Department of Neighborhoods. The funding boost will be distributed into more events and materials to bring volunteers together for work parties to build Phase 2, which will add another 1.75 planted acres to the existing site. A site

FIGURE 13.3. Volunteers work together by finding something to do or someone to help. Some volunteers are trained as leaders to help others who might have questions or who wonder where they might fit in. Photo courtesy of Jonathan H. Lee.

development design group collected community input for a year by providing maps that people could draw on during every work party. The design group, which consisted of twenty-five people, compiled the ideas into a final draft design overlaid on a topographical profile of the site. A certified landscape architect will again help produce the final version for city approval. Undoubtedly, the project will go through more rounds of public media attention and attract new and eager volunteers. While continuous inflow of fresh enthusiasm into the project has provided new ideas, energy, and skills, the core members remain the heart of the project.

CHAPTER 14

Collaborative Leadership

Through our research, we learned that community food forests almost always include more than one leader. We also learned that projects are successful when leaders demonstrate patience, humility, and persistence when working together. This chapter presents the styles, forms, contexts, and considerations for collaborative leadership. It is common for different leaders to coexist in community food forest projects. While there are skills, behaviors, traits, and leadership styles that are either inherent or learned on the part of all involved in a community food forest, overall it is a process that ultimately involves change. Peter Northouse, who has written extensively on leadership, describes the process as one where an individual influences a group to achieve a common goal.[1] For community food forests, we would broaden this definition to include the way a team of individuals collectively influences a community to create change.

Often, the idea of a food forest originates with one individual. Through subsequent phases, a team is assembled, which then works together to influence a much larger group of people working toward common goals of establishing, maintaining, and using a community food forest. Throughout the process, leadership is present in various forms, each requiring different skills and abilities.

The evolution of leadership in community food forests across the country is a remarkable store of knowledge and a rich learning opportunity. In fact, one area we were most curious about when visiting sites was how leadership responsibilities vary and are shared. Who makes decisions and how? We found social change motivated most leaders. They question dominant ideologies and status quo. They educate themselves about alternative viewpoints. They use these insights to help shape realistic possibilities for a different future. Through their own quest to learn new ideas and alternative strategies for land use in their communities, they encourage personal transformation and social awakening in others.

ASSIGNED VERSUS EMERGENT LEADERSHIP

The community food forest process helps people work together over time to learn and act in ways that better value local opinions and initiate local social change. Leading this type of work is not easy and can play out in various ways, but community food forests have created a remarkable library of information about how we can lead informal civic governance today. While the core community food forest team is being formed, individuals may be asked to take on or be assigned leadership positions, but over time emergent leadership generally takes root and shapes the group and project. Assigned leadership is necessary in situations where particular skill sets are required or early in the process to help guide group formation, such as when people at the first meeting of the Bloomington Community Orchard had to transition quickly from discussing an idea to writing a grant. One of the lessons shared by those involved is that assigning roles and determining governance structures is important to do as early as possible. If they had not done so, it would have drastically delayed the project. To get the ball rolling in such situations, it is important to assign leadership roles based on people's traits or skills. People who do whatever it takes to finish the job, are able to tolerate frustration, and are fully willing to take risks and actively engage in problem-solving exhibit desirable leadership traits during this stage.

As groups expand and grow, participation and semiautonomy grows, too. In this setting, emerging leaders are important, because maturing community food forests do not lend well to dictated, lockstep hierarchy. This is important in creating true community, not a faction or clique. Leaders arise from self-selected tasks and, in some cases, self-determined contributions. They depend on open and expressive forms of commitment. Emergent leadership is not assigned and is difficult to predict because it forms only through acceptance and encouragement of ownership and authority by other members of the group. In other words, the right conditions and organizational culture can be created that lets leaders emerge, and they must be allowed to operate without formal assignments or even specific titles. As the complexity of growth settles in, it is typical for more than one person to emerge as a leader in a community food forest project, because a variety of tasks are necessary. As the project evolves, different types of leadership are needed. At Seattle's Beacon Food Forest, for instance, leaders found it was essential to create an open, almost free-form environment that allows volunteers to provide emergent leadership by identifying where they fit in and want to take ownership over small tasks that contribute to larger goals.

Being aware that assigned and emergent leadership can exist simultaneously, as well as engendering and supporting the culture for both, can actually help prevent negative reactions to shared leadership roles. Often people who are considered managers are assigned leadership roles. Managing and leading can overlap in community food forests, but they are two different concepts. Managing a project involves organizing people and tasks and creating structure through policies, procedures, corrective action, creative solutions, scheduling, resource allocations, and budget establishment. In contrast, community food forest leadership is more variable, as it responds to new challenges and needs for change or progress and is more about responsiveness and adaptation. Leadership concerns itself largely with creating a vision that unifies, energizes, and motivates people to work together on a community food forest and clarifies a path forward. Leadership also requires that people willingly follow, whereas management focuses on structure and compliance. Many management tasks are ongoing and may become routine in a mature community food forest. Nevertheless, they should be documented once established with notes regarding what worked well so that as members in these roles move on, the procedures and guidelines can be passed along to the next managers.

Like planning for long-term succession of plants in a food forest, change in the composition of people needs to be planned for as well, and leaders must be prepared for these fluctuations. By collaborating on leadership and management needs, more than one person will be prepared to carry on if someone needs to leave the support system's core group. Management protocols that are spelled out can help groups follow best practices that will empower others in the network to take on leadership roles and develop strengths in the domain of influencing network members, building relationships, thinking strategically, and getting things done.

SERVANT AND AUTHENTIC LEADERSHIP

Servant leadership was one style of leadership that was present at many of the community food forests we visited. People lead in this way because they are interested in serving the greater good in their community and creating change that will benefit society at large. These types of leaders have an invested interest in the people willing to follow or collaborate with them and want to empower others and help them develop their full potential as individuals, not only for the good of the community food forest, but so they will also become leaders of positive change elsewhere. The term *servant leadership* was coined by Robert K. Greenleaf in the 1970s when arguing that the United States was in a governance crisis because leadership was too authoritarian and should be based more in ethics.[2] Greenleaf credited his development of servant leadership to reading Hermann Hesse's 1930s novel *The Journey to the East*, which suggests that the porter of travelers on a mission is really the leader in disguise. In caring for the group, the servant carries the ultimate authority that allows travelers to carry on. Without their presence, the "undertaking itself seemed in some mysterious way to lose meaning."[3]

Many of the characteristics servant leaders exhibit are aligned with the ethics and authentic behavior espoused by community food forest stakeholders. As servant leaders, volunteers of a food forest feel valued and connected to the project at a deeper level. Helping the food forest grow along with its associated community is a journey all on its own. This meaningful connection to the cause is dependent upon an emergent culture. Engagement is nurtured and developed by championing the personal capacities of others. The altruism of a servant leader guides the way through shared power, but this can confuse some people. The concept is better understood when the results become evident in the growth of others or when community conditions improve in response to the support of a servant leader.

People who lead this way often are good at listening to others. Through clear communication based in authenticity, they persuade others to make changes rather than using coercion. They exhibit empathy, care about the well-being of others, strongly support volunteerism, and are committed to helping others grow and take part in decision making. These leaders often have a high level of situational awareness about themselves, their peers, and social and political surroundings. They strive to model integrity, authenticity, and humility although, of course, they are human and occasionally falter. Inclusive and social justice–oriented cultures fostered by community food forests provide fertile ground for this type of leadership to grow, and it can be quite powerful.

Leadership at a community food forest should also be authentic. People who come across as intentionally authentic gain followers because they are honest, reliable, believable, and trustworthy. Others perceive them in this way because they can see when someone is confident, consistently self-disciplined, and comfortable enough with themselves to be open with others and establish strong personal relationships. Authentic leadership tends to develop slowly in someone, and the length of time needed to develop a community food forest lends itself, in many ways, to the development of authentic leadership among core group members who stick with a project. Critical life events, whether positive or negative, play a crucial role in the development of self-reflection, understanding, determination, and compassion that are at the heart of this style. Thus, when a person is involved with a community food forest over many years, they build a well of life experiences specific to the food forest and often assume an authentic position with a level of wisdom and confidence. Like servant leaders, authentic leaders are aware of themselves and their values, and they behave and act toward others in accordance with those values. They have a way of processing information that remains unbiased and unaltered by emotions and personal agendas, which can be very valuable beyond the formative years of a community food forest.

LEADERSHIP SKILLS AND TRAITS

When there is a large group of dedicated volunteers and multiple tasks at hand to be accomplished, leaders often self-direct based on their skills. This investment of human capital typically falls into three skill-based categories: technical, conceptual, and people. Someone who brings technical skills to a project has proficiency in a specific expertise area or ability to create a product. They can build, shape, and guide the mechanical and organizational needs of group projects. In contrast to working with things to produce an outcome, conceptually skilled leaders are better at working with ideas, producing themes, and communicating visions to others. They have a talent for sharing project meaning, creating strategic plans for accomplishing goals, and grasping the intricacies of abstract ideas.

Leaders who work best directly with people to accomplish group goals have a unique ability to understand and empathize with their neighbors, support and empower how they work, and capitalize on what motivates them in collaborative situations. They are able to resolve interpersonal issues, take the needs of others into account during decision making, and encourage individuals to be their best, stay involved, and contribute to the long-term success of a project.

In terms of community food forests, important technical skills range from knowing how to design and build a raised bed to being familiar with a computer program that can be used to develop educational signs. Other examples include designing and drawing the food forest layout and organizing, managing, and using a listserv for volunteer communication. Without question, a large group of people with a variety of technical skills is needed to create a community food forest. Identifying and publicizing the technical skills needed to make the community food forest "go" and allowing people to self-select roles based on comfort and experience can help volunteers find inspiration and pursue meaningful ways to be involved. Volunteers looking to improve technical skills by becoming involved with a food forest can learn by partnering with others who already have these skills and have stepped up to take on the associated responsibilities.

Conceptual leadership is critical in the growth and evolution of a community food forest because responsiveness and creative communication are necessary in light of the twists and turns a project will inevitably face. Leaders with strong conceptual skills can respond to emerging trends in the developmental process by drawing on their creative thinking, general reasoning competence, and uncanny memory and recall of important information in the midst of complex problem solving. Sometimes these skills are honed through work in a particular field; for others, it is simply a gift. Either way, it is helpful to have conceptually skilled people involved in community food forest leadership because they can stitch together different perspectives and strategically draw upon different cognitive strengths within the movement.

People-oriented leaders exhibit strong social judgment skills such as being able to adjust their behavior to a situation or person. They are also adept at perceiving the way attitudes, cultural differences, potential trigger points, or other potentially contentious factors may affect the group and impact interactions. Because leaders skilled in the human side are highly attuned to what others need, they can package information about community food forest change or goals in a way that relates to the intended audience and keeps things running smoothly. These are often the people who will be instrumental when making partnership contracts and addressing concerns of the public, agencies, or institutions that have specific interests or regulations. For example, when the Bloomington Community Orchard was forming, it was said of the founder that she was a great public speaker with an ability to share information passionately about the project and connect with her audience, which made people gravitate toward it. Her strong desire to bring people together was transferred to the community as a shared goal, which naturally motivated people to get involved.

In some cases, leadership is invisible. The shared vision of a community food forest exists between people rather than within one person. Its notion and promise is a force that leads people to stay active when embodied leadership is absent. The importance of creating a shared vision, therefore, cannot be overstated. It is something that everyone can understand and connect to. As Ross Gay, a poet and Bloomington Community Orchard board member, writes about a particular volunteer workday, "This diverse crowd earned blisters and sore backs as they sweated through their shirts to help to build their collective dream—to help feed their neighbors. Bloomington being a city (albeit a small one), most folks don't know all their neighbors. This common dreaming, in addition to apples and plums and figs, is the real fruit of this orchard. The orchard is a way for us to imagine how to better care for one another. How better to love one another." A vision that reflects people's hopes and desires for a better community and a place that feeds their need for connection to that which surrounds and nourishes them is a powerful force. It leads many to build and sustain a space that transforms aspirations into a physical, social, and spiritual place within their community. We see how it leads others to act and in turn become likewise catalyzed.

COLLABORATIVE COMMUNITY LEADERSHIP

The leader characteristics mentioned above all play critical roles in the emergence of collective community leadership. Apart from demonstrating specific skills, traits, behaviors, and so forth, community food forest leaders who are driven to work collaboratively should strengthen interdependence between all involved by striving to channel leadership toward mutual goals. Collaborative leaders inspire commitment and action by energizing others to create their own visions and solutions. They are like stagehands who work behind the scenes to bring a play together. People leading with collaboration in mind focus most of their energy on creating confidence and interest in taking part in the process.

For community food forests, the process has multiple steps from initiation to final stages. After getting people to act, leaders maintain a forward moving process by encouraging participation, celebrating small victories, and continuously setting goals. Staying committed to a process that brings about change is how collaborative leaders work. They cast a wide net to bring a variety of stakeholders into the process and in doing so involve a broad spectrum of interests. This undoubtedly takes work and thick skin, but in the long run it ensures a stronger foundation for the project because it is addressed together rather than avoided or co-opted by one group or another.

Just as there are different levels of stakeholders involved in a community food forest project, there are levels of collaborative community leadership. How the core group shepherds the project via volunteer coordination and administration is one form. Like tribal elders, they entertain and discuss different options for a collective path forward. To continue with the tribal metaphor, there are basic behaviors elders exhibit that make collaborative community leadership easier, such as upholding "family" values when interacting with the tribe, demonstrating mutual respect, voicing appreciation and gratitude, demonstrating reciprocity, and building trust through transparency and patience. These are necessary and set the tone for tribe members to identify with and feel connected to a family system that works on their behalf and that they feel compelled to honor through loyalty and dedication.

The other form is shared leadership, which reaches beyond inner circles to engage agencies, foundations, and organizations to meet community food forest goals. The core group starts working together because they are driven by a shared belief in the cause, but the next step in collaborative community leadership is to assess group values and engage group members. If values can be decided upon in the beginning, they then become a cornerstone in the collaborative decision-making process, which in turn fosters invisible leadership. When there is a hard decision to make or it seems difficult to collaborate because of differing opinions, referring to established values will lead the group to the right decision that is aligned with their mission.

When visiting with community food forest leaders across the country, we observed that stories, themes, and elements of collaborative community leadership were always present whether recognized or not. Some groups had reached a higher level of consciousness and nativization when it came to collaboration, and it was apparent this was a key to success. In many cases, it was evident in the form of camaraderie, how people looked to one another to respond to a particular question, deferring to those whose strengths played to that insight. Trust was evident and the process largely one that recounted the good and bad and formative and summative moments that led to critical investments in social as well as cultural capital.

A group culture rooted in ample social capital is what most people are looking for when they become interested in community food forests. At several sites, people likened the social experience to the evolution and growth of the food forest itself. People who are not willing to trust the process will weed themselves out. Because there are generally no requirements to participate or repercussions when one leaves, the spirit of community is defined by those who work to define it. This is a right and privilege associated with community food forests that some are willing to take on. By doing so, room is made for and by people who are excited to work together to find new and better paths forward.

In *Tribal Leadership* the authors observe that high functioning and collaborative working groups ask themselves what they call the Big Four Questions, which seem highly relevant to community food forests.[4] The questions are:

- What's working well?
- What's not working?
- What can we do to make the things that aren't working work?
- Is there anything else?

These questions might seem simple, but it is surprising how hard they are to ask, let alone answer, when the group culture prevents moving beyond power struggles and blame. Systems take time to develop to where they can stay intact and absorb such questions without falling apart. Trust the process and stay focused on the values. Another tool that has utility, given group dynamics observed across the community food forests, is the Collective Leadership Compass designed by Petra Kuenkel, the executive director of the Collective Leadership Institute. It is described in detail in *The Art of Leading Collectively*.[5]

She poses questions throughout the book that fit well when thinking about collaborative leadership using the social planning zones we described earlier. The questions can help you and others develop your own collaborative approach when leading a community food forest initiative and document personal patterns, as well as help guide the group to collectively spell out six aspects of the compass that create, prepare, and nurture an initiative so that collective governance can blossom. These six aspects are engagement, future possibilities, innovation, humanity, collective intelligence, and wholeness. If your group needs to get past barriers and obstacles in the way of creating a dynamic yet stable social system around your community food forest, this multifaceted compass is a good way to identify a direction forward.

Another form of collaborative leadership has to do with larger social systems that extend beyond the core group and other active participants to all zones of participation. Collaboratively, leadership of a community food forest across all stakeholders will take an assessment not only of values for each stakeholder, but also of their needs and limitations. Assessing a particular organization's capacity for handling some part of the project typically makes for easier collaboration because realistic expectations are created and workloads distributed. For example, site maintenance at the Mesa Harmony Garden in Santa Barbara, California, is led through collaborative activities with other stakeholder groups such as the Master Gardeners and the California Rare Fruit Growers. Rather than one person overseeing all skill-specific maintenance, the partnership among these three stakeholders lends itself to shared site maintenance, which benefits each individually. The Master Gardeners and rare fruit growers train others on particular maintenance skills, provide workshops throughout the growing season, and share values for food production. The Master Gardeners are able to promote and teach about gardening while fulfilling volunteer hours. The rare fruit growers gain access to a site where rare fruits can be grown and are able to share their passion by educating others.

CHAPTER 15

The Bloomington Community Orchard

The people and organizations who are interested in public food production oftentimes unsuspectingly find each other. No matter where someone is thinking about establishing a community food forest, chances are very good that unknown advocates and partners are right around the corner. Our study of community food forests across the country shows that with a little serendipity and a few small steps to spread the word, it is often easy to inspire your neighbors and join forces with people who are thinking along similar lines and willing to work with you toward a common goal. So even if your food forest proposal is initially considered controversial or seemingly falls on deaf ears, it is well worth the time and energy to stick with it and shop around, because there are bound to be collaborators nearby. The power of networking with them can significantly improve food availability and nutrition, as well as profoundly contribute to community growth and well-being.

Take the story of a community food forest in Bloomington, Indiana, where small steps led to big accomplishments when people with shared interests came together.

ABOUT BLOOMINGTON

Bloomington is located in Monroe County, Indiana, about fifty miles south of Indianapolis. It was founded as a settlement in 1816, incorporated as a city in 1818, and became home to Indiana University in 1820. Bloomington is often called the "Gateway to Scenic Southern Indiana" because the southern section of the state is known for its distinctive rolling forested landscape, which was once largely occupied by eroded and abandoned farmland until the state reforested large tracts of land and created lakes that would act as reservoirs for places such as Bloomington. Many of Bloomington's community members actively address local issues of food security with grassroots solutions, and the university is a hotbed of innovative ideas

Fast Facts
OWNER: **City of Bloomington Parks and Recreation Department**
LOCATION: **Winslow Woods Park, 2120 S. Highland Ave., Bloomington, IN 47401**
YEAR PLANNING STARTED: **2009**
YEAR ESTABLISHED: **2010**
SIZE: **0.8 acres**

circulating among a large mix of academics and professionals and has been the main source of the city's growth.

IDEA GENERATION

The idea for a community orchard in Bloomington was born out of the experiences and innovation of Amy Countryman, who was a student in the school of public and environmental affairs at Indiana University. Amy and her family were small-scale organic farmers who occasionally struggled to put food on the table. While a student, she spent a summer working at a friend's apple orchard in Davies County, Indiana, and was in awe of the quantity of healthy food produced by just one fruit tree. She was also surprised that much of it was left on the ground to rot. Amy began taking some of this fruit home and was inspired by the fact that she could help feed her family with the apples, peaches, and pears that would otherwise have gone to waste.

Amy's curiosity and experience at the orchard took her down a path that included questions about why trees in public spaces were not doing more than providing shade. At the time, she was taking an urban forestry class at Indiana University taught by Burney Fischer. Burney was once the director of Indiana's division of forestry and the state forester before becoming a university professor. He played an important role in the university's designation as a Tree Campus USA. Through the class, Amy and Burney leveraged their experience and interest and began discussing possibilities for increasing the number of fruit trees in Bloomington's public spaces.

Amy was required to do a senior research project as part of the course, and she chose to frame her project around the idea of an edible urban park called Bloomington Community Orchard (BCO). She began attending Bloomington Tree Commission meetings and embarked on an analysis of city trees to determine if any were fruiting species. She discovered that fewer than 1 percent produced fruits or nuts, and those were mostly mulberry and walnut trees. Since there were some fruiting species, she inquired about whether and where more could be planted.

In general, however, Bloomington was not interested in her suggestion to plant trees in the grassy curb strips between streets and sidewalks. Much of this aversion was due to the risks of fallen fruit and attracting bees in pedestrian areas and the fact that homeowners would need to learn new and different forms of maintenance. Rather than stopping there, Amy persisted and contacted the parks and recreation department. There she received a more positive reception, and her idea seemed to align with the department's new objective of reducing maintenance and mowing costs in parks.

Amy's original vision for the orchard was to plant it along the B-Line Trail, which follows railroad tracks in the city, but she abandoned the idea because the soil next to the tracks was quite contaminated. She also spent time researching whether there were models from other communities that could be followed, and although she found several examples, none existed where fruit trees were publicly owned, supported by the city, and managed by volunteers.

In 2009, she turned in her thesis proposal, which suggested that a publicly accessible site be selected in the heart of downtown and that neighborhoods all over the city allow for smaller pocket orchards to build community, contribute to the food supply, and sequester carbon.

Amy originally had no intentions of approaching Bloomington about her idea for

FIGURE 15.1. Locating fruit trees in a central area, such as a food forest within a park, can provide public benefits while minimizing risk for those allergic to stinging insects. The designated area for the fruit-producing plants can be labeled and avoided if needed. The red ball hanging from the limbs of this fruit tree is covered in a sticky substance that traps small pest insects.

BCO, but ultimately decided to present a copy of her thesis to city representatives just in case someone might be interested. As it turned out, Lee Huss, Bloomington's urban forester, was keen on the idea. Lee had also helped advise Amy on her thesis project and knew Burney, too. Other community members had suggested planting public fruit trees in the past, so Lee was aware of local interest. Now with Amy's work in hand, the timing felt right. The idea of creating an orchard in a public park presented a way of meeting community demand for fruit trees, while avoiding the liability and maintenance issues of planting them along streets. It would also cut down on mowing costs for the park, which happened to be something parks and recreation had been pondering for some time.

BUILDING SUPPORT AND FORMING PARTNERSHIPS

Bloomington's parks and recreation department houses the forestry division, which is responsible for the city's urban canopy and proudly maintains its status as a Tree City USA. In January 2010, Lee contacted Amy with an offer of $2,000 from the department and a location near the Willie Streeter Community Gardens in Winslow Woods Park. The park is home to a large baseball field, basketball court, nature trail, and playground and has a large shelter for community events. Bloomington's largest community garden, the Willie Streeter Community Gardens, has been at the park since 1984. The garden hosts conventional and organic garden plots and raised beds for a total of 178 spaces.

Burney was part of a network of local government officials and introduced the community orchard idea through conversations with other community members. Amy did not expect the response she received from the parks and recreation department, but she immediately reached out to close friends, colleagues, and scholars at the university to find allies who would be interested in helping her. She also asked those who were curious to attend and provide support at the first public meeting she organized. This meeting set initial support in motion for BCO and opened the door for a network of like-minded people.

INSTALLATION AND COMMUNITY ENGAGEMENT

In February 2010, Amy organized a public community meeting to present and discuss the idea of food forest in Bloomington. Some who attended heard about it through word-of-mouth; others read the announcement in the local newspaper. She also spread the word at the local farmers market and at the university. Remarkably, more than 100 people attended the meeting, during which people who wanted to be involved organically organized into teams to cover tasks such as cultivar selection, administration, and orchard design. Some of the attendees became members who have served on the board of directors and are still involved more than six years later.

Agency personnel were also present at the meeting. They were interested in how the food forest would align with agency responsibilities. A cooperative services agreement, which is needed for a partnership with parks and recreation before any support can be given to a project, was initiated at the meeting, and Lee was chosen to be the liaison between the town of Bloomington and community members. Although the agreement was initiated in that moment, it would not be completed until months later. As the organizers admitted, it was all moving quickly, and everyone was learning and figuring things out as they went along.

The months that followed the first meeting were a whirlwind aptly described as learning to walk while already walking. Immediately after the initial meeting, the administration group was tasked with finding funding for the project. They were able to secure $400 from Bloomington for a deer fence. A ten- to fourteen-foot-tall fence would be needed to minimize deer presence in the orchard. The group was looking for other opportunities when a national community orchard contest was announced. The grant required a proposal that described the community's dream orchard. The opportunity was perfect, and the timing was short, so the team got to work.

The Fruit Tree Planting Foundation (FTPF) partnered with Edy's Ice Cream in March 2010 to celebrate the launch of the company's Antioxidant Fruit Bar flavors (pomegranate and acai blueberry) and encourage healthy eating. Edy's Fruit Bars promoted the "Communities Take Root" program that year. It promised to plant dozens of public fruit tree orchards throughout the United States in deserving communities and would pick one deserving municipality each month from April to August. In addition to the fruit trees, FTPF would provide additional materials and support. They also promised help with on-site design, horticultural workshops, and aftercare training and manuals, along with subsidized deer fencing and drip irrigation if needed. The public could voice their opinion by voting online for the community of their choice.

The aspiring BCO team wrote a proposal for the grant for the first fruit trees, and Bloomington community members showed their support by rallying people to vote for the proposed orchard project. The contest was a testament to the far-reaching network Bloomington has access to, particularly because of the university and its alumni who share a sense of pride in the town. People would later express that by voting and encouraging others to do so, they felt as if they were participating in the community project from afar. In light of the swell of support, the group doubled down on the cause. Thirty-four people split into operations (both cultivar selection and design), finance, web, and administration teams and met in April 2010, feeling confident they would win.

The project was quickly gaining momentum. The web team worked on developing a beautiful website, which provided a virtual means for learning about the project and encouraging a vote in its favor. By the end of April, the tally saw the project take second place; they were announced as winners in May 2010 and soon received sixty fruit trees.

Looking back, the leaders viewed this process as extremely important in the development of community buy-in for the project. On the ground, site preparation had already started in April in anticipation of planting if they won. A landscape architect interested in edibles and a parks and recreation employee advised the group that site preparation should begin as soon as possible, ideally at least a year before planting.

As it was, after winning the grant, BCO could only push off planting until October, so they immediately initiated soil testing and began site preparations. The first community workday took place in May to mark off trails and mix sawdust into the large load of manure Bloomington deposited at the site. Parks and recreation employees assisted in grading and applying phosphorus to the soil and also helped install crushed gravel walking paths to comply with the Americans with Disabilities Act. The summer of 2010 was a dry year for Bloomington, and the only irrigation source available at the burgeoning community food forest was about 500 feet away at the community garden. Although preparing the soil was labor intensive and difficult, it created a bonding experience and helped build community, which would pay off down the road.

The core group of volunteers dedicated countless hours to the project to get things into place. When members were creating bylaws during one of the administration meetings, they realized they would need a board of directors. Amy pointed out that it could begin with those seated at the table, since they were continuously present and devoted to project development. The seats were filled with professionals, local government personnel, IU professors, and local artisans.

However, none had served on a board previously and were taken aback by the idea. Many did not quite feel qualified for taking it on nor did they fully understand what being a board member in such an organization would entail. Nevertheless, they all felt passionate about what Amy outlined in her proposal, particularly that food should be freely available to all and that the community

Adding Art to the Scene

Adding art to a garden increases its cultural value and creates a reflective space. Amy Brier is a Bloomington artist who works with local limestone for her sculpting medium and advocates for its cultural and geological heritage in Indiana. She was invited to give a workshop on how to make limestone benches with the BCO volunteers, and now those pieces populate the edge of the inner circular area of the food forest where volunteers sit, take a break, and reflect.

An art-collecting couple donated this sculpture that sits on a limestone bench, which also increases site aesthetics. Bringing in local professionals to share their skills with regular volunteers is a way to create events that leverage outer zones of the social system to benefit volunteers from inner zones while also developing the food forest.

FIGURE 15.2. Donated art complements other aesthetic elements of this food forest.

should partner in that endeavor. So the group agreed to move forward as the initial board, passing bylaws and encouraging the community to nominate additional board members. In the end, two more were added, and BCO took another step toward full and formal existence.

Meetings every two weeks during the first year helped build unity among core members. As these meetings progressed from discussing the enchanting vision of a community orchard to taking action and making decisions that would affect outcomes on the ground, discussions sometimes became contentious. Core members were passionate, and many had strong opinions on what needed to happen. Some initial supporters even left because they felt the timeline of the grant-supported project was unrealistic. Though the period was tumultuous, it made the group stronger rather than destroying them. It helped that they stuck to the group formation process, commonly referred to as "forming, storming, norming, and performing." Even when meetings stretched on for hours, the group talked through concerns in an effort to reach consensus, and common ground was always achieved. In the end, the result was a tightknit team that could withstand the discomfort of conflict. They went through growing pains together and emerged as a family who nurtured the project as their connection to each other continued to grow.

Through the month of August, the core group rallied their far-flung network of Bloomington supporters yet again through blog posts, social media, and local media to cast their votes in favor of the community project in another vote-based grant contest. A few days before their first planting of Edy's sponsored fruit trees in October, they learned they had won the "50 States for Good" grant from Tom's of Maine for $20,000. This grant would allow them to buy tools, build an eco-friendly shed, hire an intern, and plant more fruit trees.

Both of the grants they received hinged on the swell of support from their social capital investments that reached well beyond the relatively small population of Bloomington. At the same time, BCO also developed and nurtured local partnerships and garnered support that would significantly elevate their impact on food security and networking in Bloomington. Organizations working on food security, local food systems, social justice, and education issues in the Bloomington area would all become partners with BCO as it developed over the years. These included Hoosier Hills Food Bank, Mother Hubbard's Cupboard, Habitat for Humanity, Bloomingfoods, the Lotus Education and Arts Foundation, the Bloomington Food Security Council, and Transition Bloomington.

Mother Hubbard's Cupboard (MHC) is a food pantry as well as an educational resource for people interested in gardening and local nutrition. They provide BCO with a distribution site for surplus produce and have collaborated on grant applications and fruit and nut tree planting throughout the city and even on their own property. In 2012, the year BCO became an official 501(c)(3) nonprofit, they partnered with MHC and the city to plant fruit and nut trees in Reverend Butler Park and Crestmont Community Gardens where MHC maintains public gardening plots.

Hoosier Hills Food Bank partners with local farmers and works with community volunteers to collect excess produce and distribute it to their member agencies, which serve an estimated 7,500 people each week and 25,800 individuals annually. In addition to gleaning, they started their own organic garden in 2013 to supply the fruits and vegetables they most needed at the food bank. In October 2014, BCO donated five apple trees to the garden to ensure a steady supply of apples for their programs. They also share excess fruit from the food forest whenever it is available.

Collaborating with these organizations along with various schools, community groups, the university, and other community gardens allowed BCO to tie into their work and draw inspiration from their encouragement and collegiality. It amplified BCO's exposure in the community and drew attention and volunteers to the community food forest. Collaboration also increased grant opportunities. This paid off when the Alliance of Community Trees selected BCO for funding, noting that their grant was competitive because it included partnerships with other community organizations.

Funding agencies tend to support partnerships that are functionally diverse and spread resources throughout the community with organizations that are tied together under similar themes and act together in accomplishing an overall goal. This helps build rapport and trust with one organization, but expands their reach to a wider network. In return for having a well-run organization, BCO's connection to local agencies, strong partnerships, and diverse work outside was able to build financial capital through social and political gains. Making

FIGURE 15.3. The center of the Bloomington Community Orchard is filled with strawberries and asparagus. Espalier fruit trees demonstrate an alternative growing technique that makes it easy to harvest fruit.

inroads with key partners who are not doing exactly the same thing but who carry out work that can expand the reach of a community food forest is a key strategy for success.

CONTINUATION AND FUTURE VISIONS

The group continued to work on developing the orchard, their vision, and community values. They also focused on creating partnerships with organizations that were like-minded to create a support network for harvesting, distribution, outreach, and building an orchard community that valued being stewards of the land and each other. They did not want to expand their orchard too rapidly because they had the foresight to know their focus in the first few years should be on nurturing the trees and partnerships that would educate the public on how to care for and harvest from the trees once they were ready.

In preparation for growth, BCO decided to hold an "ask" brunch in fall 2011. Instead of inviting people to support the organization financially, the board asked community members to take on leadership positions in order to help grow the orchard's network and impact. Shortly thereafter, BCO membership and its board of directors nearly doubled, which was good because 2012 was a pivotal year. BCO achieved 501(c)(3) nonprofit status, passed articles of incorporation and bylaws with community approval, developed a transparency policy, established an official partnership with the city of Bloomington parks and recreation department, outlined a three-year strategic plan, and formalized an organizational structure.

During the same time, BCO also received funding from a variety of sources to expand the original orchard, collaborate with other organizations, offer educational classes, and hold workdays for site maintenance. They partnered with the city and Mother Hubbard's Cupboard to receive a $4,000 grant from the Arbor Day Foundation's ACTrees to plant fruit and nut trees at Butler and Crescent Parks as well as espalier pears at the orchard. Additionally, they established a partnership with Hoosier Hills Food Bank and received $2,000 from the Sunrise Rotary Club to build kiosks and install signs. To raise additional individual donations, they successfully implemented their first direct-mailing campaign and fund-raising breakfast. In an effort to connect and exchange information with other public orchards and food forests, they met with representatives from the Portland Fruit Tree Project (PFTP) and Beacon Food Forest.

The PFTP has worked with community orchards, food education, and tree care and harvesting programs since 2007, and BCO was able to gather helpful information about success and lessons learned in Portland, such as finding a community sponsor for each orchard site, ways to teach new orchard stewardship skills at "learn and work" days (and immediately put them into practice), and creating a stewardship program where volunteers sign up for monthly watering shifts and workdays for a year.

A community forum also generated suggestions for BCO to work on, such as increasing signage, social media, blog updates, and art. Signs could explain things like the fence, which was installed to exclude deer, not people, or describe how to pick fruit without damaging the trees as well as inform visitors which trees were ready for harvest. The community also wanted more photo updates on social media to promote content sharing, a blog, and an electronic forum for sharing ideas. A call for more artists to collaborate and enrich the space was also requested. Various art installations have since been created at the orchard, and other community suggestions have been met.

By the end of 2012, the organization had accomplished about 18 percent of the three-year strategic plan they had created, which outlined eleven outcomes. The first and primary BCO objective was to be a model for how perennial food production supports self-sufficiency, local resilience, and sustainability. The organization was well on its way to accomplishing such objectives and many more, and the board of directors and governance structure continued to evolve as well. They utilized *Robert's Rules of Order* (a book on parliamentary procedure, bylaws, conducting meetings, and taking minutes) to run their nonprofit with transparency, order, and inclusive community engagement. The board also held an annual meeting at the end of every year with community forums to listen to suggestions from neighbors and partners for the upcoming year's agenda.

One successful strategy of many food forests that BCO also adopted was creating seasonal festivals over the years to offer another way for community members to check out the orchard and increase interest. During the winter, there is the Hibernation Celebration where everyone enjoys fruit pies and hot cocoa during the annual meeting. Spring brings an annual planting day where new additions to the orchard are installed with community help. Toward the end of July, there is a Harvest Festival for fruits such as blackberries and blueberries. But the biggest festival is the Cider Fest associated with the fall harvest of apples, where a hand press is available for everyone to use to try fresh-pressed apple juice and cider from their own picking. Music, a scavenger hunt through the orchard, and lots of events for kids make the festival a family event, and attendance increases every year. Each fall, BCO also gives away fruit trees and provides planting demonstrations for community members.

BCO's simply stated mission of "dream, build, and share an orchard community" proved highly effective. A major part of their success came through partnerships with multiple local organizations that helped guide their mission and expand their reach. BCO demonstrated that an organization can accomplish more through strong collaboration with others. Their mission and vision were shared throughout the community, and their support system yielded more and more, like their growing orchard.

And the orchard was no longer just an orchard either. There were understory berry species and ground cover. In 2013, a new Junior Stewards (youth education) program planted herb beds with children from the nearby Montessori elementary school. Each child also took home a plant to care for and learn from.

Through other collaborations, BCO continued to provide free fruit trees and berry bushes to community members in need. National Food Day was celebrated in collaboration with the Bloomington Food Policy Council by giving away seventy-five trees and offering hands-on tree planting demonstrations. This was made possible through an Alliance for Community Trees grant. Partnering with the Food Policy Council was a direct response to suggestions during the 2012 community forum, which showed the community that their input had been valued and acted upon. The orchard had stellar attendance at workdays and family festivals in 2013 and engagement with Indiana University students increased.

The operations team also started working on a long-range orchard maintenance plan, a reference manual to share with other communities developing public orchards, and a publicly available library collection on orcharding. Additionally, they started assessing how built infrastructure at the orchard could be used to support demonstrations of sustainable technology. They would eventually use the roof of the shed for water harvesting, and they aspire to install solar power and start collecting data to provide insights on what cultivars do best in the region. It was through critical reflection and creative programming and partnerships that BCO grew and began addressing broader educational needs in their community. In 2014, the orchard was awarded $10,540 in grants, and highest costs shifted to education and outreach as they decided to cosponsor a free community orcharding class series with the parks and recreation department. Creating free classes not only matched BCO's ethos of free fruit for all, but it assisted interested community members in caring for trees planted throughout the city.

More fruit trees were installed in 2014 with yet another grant from ACTrees and educational opportunities increased for youth through the Lotus Blossom Bazaar at which Junior Stewards and middle-school children learned math lessons related to the garden and were given seed packets to take home. Classes were filled weekly from summer to fall by groups

Using Food Forests to Revive Heritage Plants

Community food forests are used in creative ways to ensure local varieties of plants are stored and available for propagating regionally acclimated plants. In Bloomington, the community food forest was used to revive a heritage species. A few Marshall strawberry plants acquired from food artist Leah Gauthier grow in one special area of the orchard.

The Marshall strawberry, a very tasty variety introduced to the United States in the late 1800s, was once commonly grown in the northwestern United States and was all but forgotten due to poor transportability and low yields. Leah saw the variety featured on a Slow Foods most endangered foods list and fell in love with it. In 2007, only a few plants remained in the USDA's Germplasm Repository in Corvallis, Oregon. While in graduate school in 2006, Leah contacted the lab and acquired a few runners by mail. She has since distributed viable offshoots from those plants to projects and gardens in Boston, New York City, New Brunswick, Maine, and BCO in Bloomington. BCO has become a repository for Marshall strawberry propagation in Bloomington. (If you are interested the Marshall strawberry, visit marshallstrawberry.com).

This is another way community food forests provide yields beyond food alone. Sites can serve as genetic repositories or sources of plant material to ensure survival of certain species or source local planting stock.

such as the Boys & Girls Clubs, Girls Inc. and a transitional housing program called The Rise! Transitional Housing. BCO also established an Indiana University student organization to improve opportunities for interns and begin planting on campus.

The orcharding class series was timed with the launch of BCO's Neighborhood Orchards Program, bringing to life more of Amy's original vision that each neighborhood include small orchards. The first location that received an edible right-of-way to plant by volunteer community members was Evergreen Village, a twelve-unit affordable sustainable housing subdivision of LEED-certified units furnished with Habitat for Humanity ReStore items. BCO partnered with Habitat for Humanity to start offering free fruit trees and berry bushes to every new Habitat homeowner. The idea was to create a "culture of abundance," and they established more than twenty trees and bushes in that first year.

A car randomly crashed into the orchard's shed in 2014. Despite the damage, the incident actually had a positive outcome in that it created an opportunity to evaluate how the orchard site could use recovery as a chance to meet growing needs. The board decided they should move the shed to make room for a covered classroom so they could continue meeting their new education focus regardless of weather conditions. The building would serve as a larger rainwater harvesting system as well as offer a protected place to hold meetings and community events. It also provided an overlook for observation of the public commons and a place for quiet reflection.

We met with the organizers of this orchard-turned-food forest as they were approaching the fifth anniversary, a year when some of the fruit trees were beginning to reach maturity and yields were increasing. In addition to plans for starting construction on the outdoor classroom in 2016, it was time to update the original strategic plan to guide future growth. Building upon the success of their neighborhood plantings and tree giveaways, the group planned to continue planting in public commons and educating citizens on how to care for and harvest from the trees. In 2015, the first campus plantings occurred on the Indiana University campus and organizers developed their own leadership team, applied for grants, fostered on-campus partnerships, and laid out maintenance and harvest schedules. BCO wants to continue this work and form student organizations at other local colleges and educational institutions as well.

BCO also has a dream of finalizing a community orcharding handbook, along with continuing fruit tree giveaways in the fall and spring. They would also like to develop a tree-tracking system and research database to provide more information about their edible urban canopy, monitor factors that contribute to successful tree growth, and shape the educational programming they offer community members. The information gathered would inform regional decisions on cultivars and growing techniques as well.

For BCO, the success of their mission is measured by the pride the community has in its orchard and the awareness of their role in a national movement that is helping communities civically engage through use of public commons to steward their environment and community by growing perennial food. What is remarkable about the BCO story is that one person sketched an idea based on personal experience. When the initial proposal did not draw support, she adapted and reached out around the community to people she knew and even some that she did not. This formative effort opened the door to an explosion of partnerships and support that were not always easy but often highly impactful. In many ways, BCO's story is a parable about the role and responsibilities of a community food forest, as well as its promise and potential. Perhaps the person standing next to us is an advocate and is interested in our project, but we will never know unless we try to find out. BCO's success tells us that we should continue striving to find more allies and use growing numbers of partnerships as stepping-stones to lead the way to meaningful change and greater chances for permanence.

CONCLUSION
Looking Back, Moving Forward

Community food forests have made their way into many of the largest metropolitan centers in the United States as well as town squares, suburban parks, and neighborhood lots. Although food forests have a positive track record on private residential properties, a decade ago the concept at the community scale seemed almost ludicrous to some. Pioneer efforts met skepticism about projects bound to end in a tragedy of the commons. Yet the stories we tell in this book depict an immense amount of collaborative and exciting work to improve lives, landscapes, and communities using public food forests. We hope these chronicles of perennial food projects continuing to unfold across the country have inspired you. This book serves as a much-needed repository for an emerging movement, and we believe the timing could not be better to help increase chances for success across the nation's ever-growing body of community food forests.

CREATIVITY AND COMMUNITY

In New York City, a 5,000-square-foot barge named Swale is in an awe-inspiring art project that floats a community food forest from one pier to another. Wherever it docks during the growing season, visitors can step onboard and forage for food while learning about the perennial polyculture practices of a community food forest. As with many groundbreaking projects, Swale is a creative work-around in a municipality that has policies against food foraging and gleaning in public parks and forested spaces. By locating Swale on the waterways, the food forest sidesteps such restrictions.

Artist Mary Mattingly designed Swale as an educational platform to bring attention to urban food issues and urge New York City lawmakers to rethink their stance on edible perennial landscapes. It is ironic, but the city department responsible for enforcing prohibitive policies actually helps fund Swale and thereby aids the bold and creative strategy in the face of its own legal constraints. Swale launched its inaugural summer 2016 tour from one of the largest food deserts in New York City and continued touring piers in 2017.

Although it receives scores of visitors every day and notable public media attention, the project is not a permanent solution in the spirit of a perennial edible food commons because of its high cost. But it is a positive step toward creating change by raising important questions about what limits the installation of fixed edible perennial landscapes. Furthermore, it is an example of how creative people use avant-garde thinking (i.e., human and cultural capital) to

CONCLUSION

enact environmental, social, and political change. Those who are willing to question the status quo and create a different future constitute an emerging class of community food forest leaders in the United States, and we were honored to learn from them and share their stories.

Youth in the United States are also energetic allies tackling public food challenges and enacting change in the places they live. In Houston, Texas, for example, students at eleven schools in the city's impoverished East End recently created a program called Green Ambassadors to address access to healthy food. They have planted more than 200 fruit trees and food gardens to transform their neighborhoods into green corridors. Their actions and goals are based in curriculum available through Project Learning Tree, a national program that creates environmental education resources for schools. This example speaks to the role younger generations are playing in creating food-producing commons that benefit communities and engender positive change. They are learning at a young age that by working together they can transform their local environment and by doing so, they are becoming the next generation of community food forest leaders.

In this book, we focused on the challenges of creating community food forests that effectively engage people and achieve intended community impacts. How do people define the ideals, concepts, and hopes that come with them? What does this development of growing and gathering in the places people live really signal? In chapter 6, we outlined how technological and agricultural advances have directed industrial and economic growth in the past 150 years. Products were increasingly made from base resources, and most of our goods and services were commodified. Human connections changed as the world became more complex, consumerism became more important than bartering or exchanging favors, and financial wealth ruled over other forms of capital.

Community food forests offer an opportunity to take part in an alternative economy, one where status and security can be achieved by investing in and exchanging diverse forms of community capitals. Participating in a community food forest strengthens the bond between people that comes from sharing resources rather than competing for them. Working the land, planting seeds, nurturing growth, and harvesting food together recaptures what many perceive as a lost link between people and food production.

Instead of trying to gain a position of power over others, people can find a sense of satisfaction through what can be accomplished by sharing power *with* others, and this dynamic is essential to the contemporary community food forest movement. Collaboration and cooperation that brings about positive community change is empowering and ignites our imagination about what else is possible when people band and act together. However, such an experience and its accomplishments take time, which is why perennial food systems are advantageous.

Stephen Hale, a founding volunteer at Bloomington Community Orchard, initially became involved as a way to meet Master Composter volunteer requirements. During an interview with a local radio station, he recalled an epiphany during his fifth year working at the food forest. He and others took a break to share homemade apple butter made with ingredients harvested from the site. It suddenly hit him that he felt a strong sense of camaraderie and community with those people, and through food, he felt more strongly connected to the space they had tended for years.

CONCLUSION

The act of stewarding a project over time is one of its most fundamental impacts. Stephen's awakening was especially attuned to the time it takes to nurture and manage a community food forest. He noted pruning or fertilizing a plant shows immediate results, but it is not until much later—sometimes a year or longer—that one really grasps what they are creating. Such realizations can be powerful, often emerging during a seemingly simple moment when you find yourself standing in the company of others that you have been working alongside. Just as it often takes several years for a fruit tree to produce its bounty, community also needs time to grow and fortify at a food forest.

Our experience suggests community food forests shape and support a different form of economy. They provide a type of banking system that accepts, exchanges, multiplies, reinvests, and directs investments in a wide range of community capital. Community food forests are part of an evolving social story, which merges our past with new knowledge, practices, technologies, and applications. Permanence is the goal, but whether community food forests persist or not, what matters is they are shifting and shaping people's views of the society they want to create and live in. In that regard, we believe that both the long-lived Dr. George Washington Carver Edible Park and the short-lived Hazelwood Food Forest can stake a claim in being a fundamental part of change.

In our current economy, unregulated loans to anyone who wants one can lead to market collapse. At the same time, overregulation means people who need a loan may not be able to get one. Understanding who has access to resources at a community food forest as well as who regulates that access is a critical distinction. Every project will need to find an approach that meets the needs of a community while balancing production capacity and demands on volunteer time. Project leaders need to be realistic about the size of their food forest and how it can best support a community. It might take time for people to understand the role they play in regulating the exchange of community capitals so that sites remain open and accessible to everyone.

There are many more community food forest projects underway than those covered in this book. It is reasonable to assume that all are grappling with the ups and downs, questions and concerns, and goals and gains that we observed and that were shared by project leaders. The resulting narrative provides fertile ground for learning and improving the experience and impact of the next generation of community food forests. This is the reason we undertook the work of gathering perspectives and experiences at sites across the country. The following section summarizes ten of the most important lessons we learned during our work. An overview of current community food forest research and developments follow. Lastly, we provide thoughts about next steps for projects in the United States and beyond, along with our beliefs about the possibilities for a meaningful legacy.

LESSONS LEARNED

Unlike community gardens, community food forests depend on perennial vegetation and thus take years before the benefits they provide come to fruition. This can be an advantage, because long establishment periods afford opportunities for thoughtful planning, reflection, adaptation, and networking. However, it can also lead to fatigue, confusion, frustration, and dismay as people wait and wonder and enthusiasm wanes.

CONCLUSION

The process of creating a community food forest and committing to long-term maintenance helps convey what it truly means to participate in local democratic processes.

As we emphasized in the introduction, food forest design is spelled out in great detail in many books, but they typically focus on biological structure and function. Little is covered when it comes to explicitly shaping food forests that sit at the intersection of perennial food system design and the deep complexities and uplifting prospects of community development. Yet people have been testing what works and what does not in the developmental stages of a community food forest, and the following ten themes are a synthesis and recapitulation of what we have gathered from the case studies presented in this book, various discussions with site leaders and participants, and our own experience.

Just. Do. It. Persevere!

Invariably, there will be obstacles along the way when creating a community food forest. One of the biggest hurdles can be land—more specifically, the long-term use of land. If the initial response to your request for property use is negative, heed the takeaway message from the Beacon Food Forest in Seattle: Do not be deterred. A strong core group needs to form and do their homework. You may find existing political structures that need change, such as outdated policies, or find flexibility within the structure to accommodate your project. Unrealized initiatives by local agencies or organizations might provide an opportunity for creative solutions or negotiations that can lead to desired outcomes for all parties involved. Stay committed to the project and show others that you are not easily giving up or going away. Work diligently to build a community food forest social system that can persevere when burnout sets in, and distribute the burden of navigating bureaucratic obstacles.

Equally important is staying flexible about the design and expectations of stakeholder and volunteer involvement. Keep the end goal in sight. In other words, plan the project with funding and community support regardless, even if it means that plans need to be adapted to fit certain regulations or community concerns.

Flexibility Is Key

Perhaps the most profound lesson of the Basalt Edible Food Park is its most simple: Being flexible and remaining optimistic works. Though it sounds a little Pollyanna-ish, the power of positive thinking influences all aspects of a project. The need for such thinking and a willingness to stay open to the unexpected and roll with the punches is nearly universal for projects that involve a diverse group of people and rely on sustained effort over time. At a minimum, developing a site on public or leased property brings an extra level of decision making into community food forest design that may delay a start date or slow progress. Such hurdles would not be a problem for a project planned for a backyard or on a private lot. Staying optimistic but flexible about changes to scheduling or original ideas will show stakeholders their concerns matter.

Engage in Dialogue

Another lesson learned from the Beacon Food Forest is that too much publicity before there are tangible results to show can be overwhelming. It is difficult and potentially frustrating to answer questions when all you can offer is a vision for the future and an explanation of challenges that have delayed on-the-ground progress. Yet publicity is also an opportunity to get the word out and build a community of interest that is willing to volunteer once approvals are granted. Media attention, especially the wealth of online sources, can mean instantaneous engagement in dialogue about your project with the greater public. Responses to articles can highlight what values resonate with the community and bring up valid questions and concerns that should be addressed in a long-term strategic plan. Although publicity may have complicated the development of the Beacon Food Forest by distracting leaders and team members, it ultimately paid off and became an integral part of the project that increased motivation and attracted global attention and support.

In addition to conversation with the public, entering into dialogue with local agencies is also crucial. Regardless of how long it takes, dialogue is necessary and can help all sides strengthen alignment with project goals and build trust.

Keep Adding Layers

At the project's inception, the core group for the Bloomington Community Orchard decided to call it an orchard because they did not intend it to be a food forest. The site design was focused on simple plantings of fruit and nut trees, but adding more and more layers—crops that would serve the community, from strawberries to asparagus, to berries and herbs—happened naturally.

The canopy species are typically the first layer installed in a food forest, so adding additional layers of vegetation as funding or volunteer time allows can enhance biological complexity at the site, attract pollinators, and offer participants new food options and educational opportunities. Many cities have opted to install community orchards rather than food forests, but there is always room for expansion. The opposite holds true as well: If it proves too difficult to maintain the subcanopy species in a food forest, plants can be removed to simplify the layout. This is an important option to keep in mind, even during the planning phase.

The Dr. George Washington Carver Edible Park was originally designed as a stacked permaculture food forest. Over time, five or six layers were successfully installed. More than 80 trees and 100 shrubs were originally planted around a boardwalk that clearly marked the main path. The canopy layer at the community food forest is mature and tree crowns have grown close together, creating a highly shaded understory (80 percent plus). Now there are approximately two layers in the main portion of the food forest—many of the canopy species are fruit and nut trees, and there is some naturally spreading substantial ground cover. Where light enters adjacent to the path, understory growth is more abundant. Part of the property has a steep slope that often gets filled with invasive species such as kudzu. Over the long term, changes like this can happen in any food forest. Thinning the forest canopy becomes necessary in later years if trees grow too thickly into each other or invasive species become a problem. A mature food forest mimics the dense canopy of a mature forest, and because of this, tending is required as in any managed forest.

Create a Governance Structure Early

Back at the Bloomington Community Orchard, a grant opportunity materialized shortly after the idea of creating the orchard arose, well in advance of mapping out a plan or drafting a mission statement. Site leaders had to organize quickly to write the grant. They held multiple meetings, working through their values, community building, and preferences for site design. In the end, the ups and downs they experienced helped form stronger bonds, and those who were skeptical self-selected out of the group, which aided progress. In hindsight, though, the group saw that if they had been able to start out more slowly and put governance structures into place early, they could have avoided some long and painful meetings. Thus, establishing at least a loose organizational structure for your community food forest project from the outset will help systematize procedures, which can pay off when it comes to efficiencies and responsiveness. The community food forest in Bloomington now has multiple committees, which makes it easy to understand division of responsibilities. There are also clear guidelines for volunteers about where and how to best utilize their strengths and who is in charge of different aspects of the project.

People Count, Volunteers Matter

The first way to keep volunteers coming back is to empower them and encourage community food forest ownership. The second is to keep work parties educational, entertaining, and fun! Leaders from across the country told us how important it was to make sure being involved with a community food forest was enjoyable and useful. When people have a good time and walk away feeling better off for having participated, they are typically apt to come back. Volunteers provide brawn, brains, and passion and can be instrumental in community food forest installation and permanence. They make up a diverse cast of characters that creates a healthy social system. Though running a volunteer program can be challenging, volunteers help reduce burnout in the core group and infuse energy into a project.

Beacon Food Forest volunteer coordinator, Jim Irby, hosted monthly volunteer orientation meetings so new participants could easily understand the organizational structure and involve themselves in ways that aligned with their interests and strengths or allowed them to learn a new skill. Barriers to taking on a leadership role were diminished by holding quarterly stewardship trainings and creating a community culture that welcomed fresh ideas and new people to monthly meetings. Additionally, volunteers were rewarded at every work party (held monthly) with a lunch and social time. Small donations ($2 to $5) were suggested, but not required, and a hearty meal of soup, vegetables, bread, drinks, and desserts were provided to celebrate growing, harvesting, and sharing food.

The leaders of the Beacon Food Forest fostered a sense of ownership by engaging volunteers on a regular basis and empowering them each workday to take on small projects that contribute to larger installation aims. People were drawn to engage out of curiosity or a sense of hope in what the project could accomplish, but ultimately, they felt invested in the food forest's success because they contributed their time, effort, and emotion toward building a shared vision. Glenn Herlihy, cofounder of the Beacon Food Forest, and others have been amazed at the diversity and sheer number of people who regularly show up to help. Community outreach includes providing information in a variety of languages to attract people from the surrounding neighborhoods.

Set Up for Success

Feed them and *make the workday* fun are important general rules of thumb when coordinating volunteers, but more specifically, a little staging can go a long way. To set up volunteers for success, it is extremely helpful to prepare the site beforehand. Tasks that can be done in advance include laying out tools, digging starter holes, flagging off planting areas, and staging and labeling planting stock. As demonstrated in Basalt, this type of advance preparation will help volunteers achieve noticeable progress by the end of the day, so they sense impact and change related to their effort. On that note, it is also very important to show appreciation for the time volunteers put into a project. Project organizers in Basalt made sure to publicly recognize their volunteers, which helped increase the efficiencies of this aspect of the work on their community food forest.

Find a Champion

Beyond volunteers, it is also critical to engage people in the community who have jobs that can benefit a community food forest through direct professional relationships. Approaching municipal employees, for example, to secure representative support for the project early on can be very powerful. A local government advocate can help immensely with funding opportunities and positioning the project as a priority in a township or city. They can also explain and anticipate concerns from different departments when writing proposals, navigating policy, and planting and maintaining a food forest. As noted in chapter 9, buy-in from aligned professionals such as a town horticulturist, landscape architect, park manager, or arborist can pay off when it comes to initiating and seeing a community food forest through to installation and beyond. In planning the Basalt Food Park, the involvement of a wildlife scientist was critical as due diligence for safety and humane purposes.

Celebrate!

The BCO in Bloomington celebrates multiple times throughout the year based on themes like cider pressing in the fall or special dinners during the winter. Annual celebrations give regular volunteers something to look forward to where they can show off the site they have been working on. Celebrations are also an ideal time for peripheral stakeholders to re-engage with the site and learn about progress and impacts. Smaller celebrations throughout the year geared more toward regular volunteers strengthen community ties, but when it comes to big events, spread the word broadly. These events are also a good opportunity to call media attention to your community food forest and generate donations or funding.

Reassess and Grow

The Dr. George Washington Carver Edible Park is one of the few we visited that offered insight into the scope and number of changes a community food forest might experience over time. The input gathered through stakeholder meetings demonstrated that a community food forest, like other natural features, experiences times of disturbance and uncertainty, and that periodic interventions are not only beneficial but most times necessary. Reassessing what has happened to date is a natural opportunity for growth and revitalization, so welcome it with positive energy.

In a natural forest, disturbances such as lightning or a heavy wind event often create gaps in the canopy that allow light to reach the

CONCLUSION

forest floor where seedlings have been waiting to sprout and thrive with newfound access to previously unavailable resources. Similarly, the partnerships and stakeholders of a food forest may also experience times of disturbance that create gaps in management, but these openings create the necessary space for new human resources to enter, rekindle energy, and spark innovative ideas that help define the next phases of growth.

CURRENT TRENDS AND FUTURE POSSIBILITIES

After our first encounter with a community food forest, we were curious to find out more about them. What we discovered was a growing but unformed movement with a unique history. Its momentum has been building for decades and, with a few earlier exceptions, physical results are now visible at multiple sites around the world. In addition to the United States, Australia, New Zealand, England, and Canada have also witnessed growth in the number of public food forests. We would not be surprised if other countries are seeing much the same. Tomas Remiarz, author of *Forest Gardening in Practice* and one of the founding members of the Permaculture Association's research advisory board, has been cataloging private, public, and commercial food forest sites throughout Europe. He is helping to start the Food Forest International Research Network, which aims to link food forest researchers around the world and develop research projects that study the ecological, economic, and social aspects of food forests.

Overall attention to sustainable urban development and urban agriculture has increased globally at a rapid pace. Much of this is part of prudent planning as the world braces for substantial growth in urban populations and seeks to address climate variability and build resilient food-producing systems. The ecological design of production systems like food forests creates opportunity for natural mechanisms, like resiliency, that help sustain environmental and social well-being in the places people live. In this next section we present trends and advancements in urban food forestry that we believe are central to future possibilities and growth of community food forests.

International Support for Urban Food Forestry

Once every twenty years, the United Nations General Assembly convenes Habitat, an international conference on housing and sustainable urban development. The goal is to renew political commitment for sustainable urban development, assess accomplishments and failures, and address new and emerging challenges. In October 2016, Habitat III took place in Quito, Ecuador.[1] An exciting result emerged for community food forests. As part of the New Urban Agenda developed and agreed to at the meeting, the Food and Agriculture Organization (FAO) of the United Nations produced *Guidelines on Urban and Peri-urban Forestry* that specifically outlines a path governments can use to eliminate policies and regulations that are barriers to urban food forestry.

The FAO recognizes that urban forests can be sources of highly nutritious food for urban dwellers and promotes urban food forestry planning for food and nutrition security. The guidelines from Habitat III encourage use of public lands (parks, schools, vacant lots, and streets) to produce food through food forests and community gardens. It also calls for an assessment of the ecological and social impacts of urban forest food production. The FAO document stands to help catapult community food forest development forward as international attention is drawn to critical food supply and environmental problems.

Urban Food Forestry and Scientific Research

While working on this book, we were able to connect with others who were also learning about how community food forests serve their communities and create and sustain healthy and productive local ecosystems. Graduate students in programs ranging from landscape architecture to natural resources management are researching community food forests and studying design and management strategies. One has focused her research on food forests in Tempe, Arizona, and Germany. The aim is to explore differences and similarities in management options as a way to synthesize global best practices. Another is studying food forests in Canada to develop criteria and indicators for monitoring contributions to ecological restoration. A third student will be studying sites from the perspective of public administration, looking at sources of funding and economic return on investment.

The growing body of research includes topics from design features that contribute to public access to quantifying yields produced by fruit and nut trees. At the time of this writing, a landscape architecture student who studied and compared the Dr. George Washington Carver Edible Park and the Beacon Food Forest is now helping to design a large-scale project, the Browns Mill Food Forest in Atlanta, Georgia. With an area of more than seven acres, this project will be comparable in scale to the Beacon Food Forest. This is just one example of how research results can directly influence future community food forests.

A leading international journal for research on urban forests and green spaces, *Urban Forestry and Urban Greening*, has dedicated a special issue (projected for publication in 2018) to research on urban food forests. The issue will be the first collection of research on urban food forestry associated with the following topics: (1) planning, (2) governance and management, (3) benefits, and (4) potential threats to human health or environmental disservices from urban food forests. A special issue among academic journals is a testament to the accomplishments of many and inspires top talent in the research field to study gaps in our knowledge and suggest ways to fill them.

Increasing Funding Opportunities

Government agencies are recognizing community food forests as urban sites for conservation and agroforestry. Some community food forests have struggled to acquire funds needed for site development, but many have found funding through local or national government dollars across a range of agencies and grant programs. The Browns Mill Food Forest is an excellent example of how funding for community food forests is increasing. The project is supported by the USDA Forest Service, the National Park Service, and the Conservation Fund, in collaboration with the city of Atlanta's office of sustainability. We expect this trend to increase in the coming decade. A national grant from the US Forest Service's urban and community forestry department was awarded to the nonprofit Inhabit Earth (formerly Earth Learning) in 2016 to support community food forest research focused on developing a database of plants that work well together in different project regions. Results will be used to create a template for town codes that can support installation. As information about community food forests grows, we will continue to post results and updates at www.communityfoodforests.com.

A Need for Formal Connection

Community food forests across the United States are not formally connected to each other in a way that easily allows shared communication of goals, accomplishments, challenges, and opportunities. Thus they are not united in what could be considered a broad social movement, although community organizing sometimes happens at the regional scale to gain support for a project. Each is taking place primarily at a local level with local players, even though some may be inspired by what is happening in other towns and cities. As the number of food forests increases in the United States and abroad, a logical evolution would be the development of an association or society or formalizing a community of practice that regularly shares information and advances knowledge.

As we conducted our research for this book, we came to realize that interproject connectivity is very limited. The value of a practitioner network is obviously powerful because it strengthens problem solving at a site and streamlines best practices for project design. Formally connecting initiatives under a common cause or a national or international association is a way to build a larger identity and expand community food forest marketing and awareness. There are extensive examples of the value of professional societies or associations, but a loosely formed guild can also play an important role. One possibility is to form a network wherein experienced and early-stage community food forest leaders and teams assemble in person at a dedicated community food forest conference, be it regional, national, or international.

Work at such an event could be geared toward collectively defining the process of community food forest development by formulating agreed-upon guidelines and mission statements. Resulting information can then be shared with others who are interested in starting a community food forest and then revisited and refined at later points in time as the movement grows and additional meetings are warranted. Another idea is to convene a community of practice as part of a national or regional community development or sustainable agriculture conference, such as an annual conference of the Arbor Day Foundation Alliance for Community Trees or the American Community Gardening Association. Either way, as community food forests grow in number, so too will the need for interproject communication.

A LEGACY OF COMMUNITY YIELDS

Community is crafted at food forests that invest substantial time and energy in maintaining social systems and positioning their work as something more than a solution to food security or a project of plant selection and landscape design. Community food forests serve as places for community renewal where social well-being is achieved through fellowship, recreation, and art. They also provide environmental benefits realized through services such as soil and water conservation, air filtration, microclimate management, and improved habitat. Sometimes they are sensationalized as Gardens of Eden that will quickly improve food security and nutrition in low-income neighborhoods and

CONCLUSION

profoundly affect environmental sustainability. It is not easy to measure changes in the societal trends or outlook that we mention throughout this book, but we believe a broader understanding of the role of community food forests in re-enlivening places and people is important. Our comprehension will continue to evolve.

Building cross-cultural community is not yet fully part of the scope at many sites, but much of this may be related to ongoing formative efforts to navigate policy, find funding, and initiate design. Many groups are actively searching for ways to include cultural diversity, but most find themselves focusing their effort on bonding instead. As more community food forests move forward, additional attention to diversity and inclusion will be important, though leaders should be careful not to force fabrications. In urban areas where people are often disconnected from their neighbors, community building in general is a positive step toward strengthening the social fabric.

Learning about community food forests was also a process of learning about the mechanisms leaders have used to assemble and excite people about a new future, which often has a great deal of political charge. Those who went directly to work on the business of creating a community food forest usually learned a great deal quickly about local politics and city values. But in many cases, delays and complications only fortified leadership and their goals, often drawing in new partners and creating bridging capital with existing volunteers. What seemed to be the greatest invigorator, though, was the thrill of imagining how the project might unfold. The spark of what is possible inspires people to wonder what else they can do in their community that would improve it but was not previously thought possible. It is also important, though, that excitement be grounded in action.

Another source of satisfaction at community food forests is seeing the immediate results of labor. It is about knowing that the result of our work will literally grow roots in the ground and continue to bring forth benefits to others for years to come. In a world of delayed results, increasingly transient lives, and an overabundance of negativity, planting fruit trees in a food forest is satisfying in ways many of us rarely get to experience, and thus has profound meaning. Those benefits will continue even if the food forest itself never bears fruit. And even if volunteers do not stay involved for the life of the food forest, they learn how to provide valuable civic and environmental services. Such an intersection is at the heart of community food forests and reflects a value proposition that sustainable, nutritious food and vibrant, diverse communities are necessary for global well-being. Through engaging with a community food forest, even only a few times, many volunteers learn how to redesign public space or their own yards, using productive ecological landscapes to improve access to nutritious food.

In terms of the community capitals, community food forests reinvest in each form, even if the project itself fails. People become involved with community food forests for a range of motivations from political to environmental to personal. Some are conservationists interested in land rehabilitation, and others are intrigued by the idea of walking along pathways lined with trees bearing ripe fruit. There are those who are politically motivated about food distribution systems and believe that foraging knowledge is fundamental, whereas some may simply be interested in connecting with positive-minded people and a project that is grounded in hope and resilience.

CONCLUSION

When we visited the Beacon Food Forest, one important topic of conversation was why people participate. Most people want to get their hands into the earth and enjoy the satisfaction of putting their whole body to work on something tangible. Many considered it a concrete action that gives back to the land and also an opportunity to create and be part of something larger. It was a process of shaping a new way of thinking about and nurturing community, participating in a different form of economy, and regaining skills to meet basic human need.

When a community food forest is thoroughly planned out in ways described in this book, the community capital gains can continue to multiply. At the time of this writing, there are more than seventy community food forests in the United States. We have told the stories of just a fraction of them, but we hope they illustrate that community food forests can provide all types of yields. In addition to edible goods, they grow learning experiences, communities, habitat, change, and perhaps most important, innovation and inspiration. Judging their potential only by the quantity of food yields they can produce is missing a large part of the harvest. This is the promise of community food forests in the United States, and in this promise, we find action and impact.

Acknowledgments

In these acknowledgments, it is quite possible we have missed thanking some people by name, but we have not forgotten you. This book would not be possible without the contributions of the many, many people who have been involved and continue to be. Multiple colleagues and friends engaged in important conversations, listening, and support that helped this work evolve: We give our deepest thanks to all of you. We would also like to acknowledge all of the people pushing community food forests and local food systems forward, making projects happen, shaping dialogue, and creating change in communities. Our future needs you and we are honored to be a part of your work.

First, we want to express our gratitude to the Chelsea Green team for believing in us, particularly Makenna Goodman and publisher Margo Baldwin for encouraging and accepting our proposal, and Fern Marshall Bradley for all the hours, patience, and heart to push this book across the finish line.

We would also like to thank Virginia Tech for supporting this type of research; especially VtEngage for sponsoring the Hale-Y Community Garden Food Forest, which provided a valuable learning ground. We especially thank Jenny Schwanke and Arlean Lambert for all their support with the food forest and their never-ending positivity. Catherine would particularly like to thank her dissertation committee members for their patience and guidance: Kim Nielwolny, James Chamberlain, and Paul Kelsch. We also thank Jay Sullivan at Virginia Tech and Janaki Alavalapati, now at Auburn University, for leading departments that value our work. Additional thanks to William Serge for the information and drawings from his landscape architecture thesis on the Growing Goodwill Garden in Roanoke, Virginia.

Over the span of the last four years, over sixty people participated in interviews, focus groups, and site tours. They shared photos, stories, conversations, and new discoveries; they offered their knowledge, insights, and encouragement. They made space for presentations to spread the word and helped connect us with people who provided access to the world of community food forests, including their perspectives and experiences. We have a deep appreciation for all of the leaders and participants at the community food forests we visited, who are the pioneers making this work visible.

Special thanks to those who helped arrange meetings and tours at sites, as well as providing information and feedback on the stories we tell in the book: Andy and Sean in Arcata, California; Jonathan, Samantha, Al, Lindsey, and Darcel in Asheville, North Carolina; Stephanie, Lisa, and Jimmy in Basalt, Colorado; Amy in Bloomington, Indiana; Allison and Daniel in Boston, Massachusetts; Chris in Davenport, Iowa; Lincoln, Kim, Alex, and Jeannie in Greenbelt, Maryland; Jessica, Caroline, and

ACKNOWLEDGMENTS

Christopher in Helena, Montana; Fred in Iowa City, Iowa; Taylor in Johnson City, Tennessee; Adam and Terry in Lincoln, Nebraska; Stephen and Alicia in Merriam, Kansas; Therese and January (may your spirit continue to bloom in the food forest) in Pasadena, California; Marlon, Maritza, and Robyn in Philadelphia, Pennsylvania; Seth and Juliette in Pittsburgh, Pennsylvania; Angela in Portland, Oregon: Kate and Jamie in Providence, Rhode Island; Sherman in Richmond, California; Keith and Paul in San Diego, California; Kevin in San Francisco, California; Hugh and Larry in Santa Barbara, California; Glenn, Jonathan, Peter W., Peter L., and Matthew in Seattle, Washington; Frank and Jason in Syracuse, New York; and Rebeka, Kathy, Branda, and Steve in Troy, New York.

Thanks to all of the Indiegogo campaign supporters who believed in and encouraged this project, particularly to those who purchased advance copies and gifted them to others.

Special thanks to Jonathan Lee of SubtleDream.com for his photos; Tomas Remiarz for insights and starting a network to connect the many doing research on this topic; and Dave Jacke for sharing his knowledge during conversations and the forest garden intensive.

Thanks also to our colleagues at USDA's National Agroforestry Center, the Association for Temperate Agroforestry, and University of Missouri's Center for Agroforestry who have helped promote and expand this work (Susan Stein, Kate McFarland, Gary Bentrup, Rich Straight, Mike Gold, Gregory Mori-Ormsby, and Erik Hagan).

We appreciate those who provided space for presentations and promotion, especially the following invitations: Diego Footer at Permaculture One; Grayson Landcare, the first rural community food forest project we know of; the Arizona Community Tree Alliance; and Kathy Sheehan, USFS Urban and Community Forestry Program.

Catherine would like to specifically thank Sandra Bukowski-Peverly and William Peverly and Carol and Elliott Dworin for supportive and worry-free places to write and simply everything you have done; Ama for your daily reminder to get outside, walk, and laugh on serious days; Jonathan Dworin for helping out whenever asked, road trips, and sticking through it all. Huge thanks to Susan Perry for always being willing to read, review, and provide great editing during many renditions. To all my loved ones who listened during the frustrating times and celebrated the good ones. John simply thanks all of his family, friends, and colleagues for their encouragement, support, and conviviality.

APPENDIX

Goals, Visions, and Mission Statements Associated with Community Food Forests

Mission, vision, and goal statements epitomize the aspirations of a group. Here, we've assembled statements from community food forests across the country. These statements are good examples of the patterns of change sought through the development of community food forests. You may find examples of language that you can adapt for your own project.

Dr. George Washington Carver Edible Park, Asheville, NC

"Combining elements of park, community garden, and permaculture garden, the Bountiful Cities Project introduces edible, public, open spaces to urban and suburban areas. Perennial fruits and vegetables, along with other useful vegetation, are combined with swings, picnic tables, and park benches to create a low maintenance garden paradise in the city. We are urging other cities to follow our lead and establish useful vegetation in public space."

Source: 1997 newsletter of City Seeds, a now defunct nonprofit organization that in partnership with the City of Asheville Parks and Recreation Department supported the Bountiful Cities project, which developed the Dr. George Washington Carver Edible Park.

18th and Rhode Island Permaculture Garden, San Francisco, California

The mission of the community food forest is "to improve local food security. Serve as a home for rare and productive flora—a living library of genetic stock. An outdoor classroom. A place to engage, relax, enjoy. A site for the neighborhood and community at large."

Source: https://www.facebook.com/18thAndRhodeIslandGarden/

Beacon Food Forest, Seattle, Washington

"The goal of the Beacon Food Forest is to design, plant, and grow an edible urban forest garden that inspires our community to gather together, grow our own food, and rehabilitate our local ecosystem.

"Join us to improve public health by regenerating our public land into an edible forest ecosystem. We work to reduce agricultural climate impact, improve our local food security, provide educational opportunities, and celebrate growing food for the benefit of all species."

Source: http://beaconfoodforest.org/

APPENDIX

Festival Beach Food Forest, Austin, Texas

"We intend to transform three acres of a ninety-acre park into an edible forest garden where visitors can openly forage and enjoy fresh food on the shores of Lady Bird Lake. Fruits, nuts, vegetables and herbs will now inhabit the city-owned land.

"This project was first envisioned by community activists from the East Feast Coalition who recognized the power of community members connecting, growing, and celebrating together. Grounded in their neighborhoods' heritage and inspired by the possibility of renewing relationships with the land and with each other, they gathered neighbors and allies in support of edible landscaping.

"The Festival Beach Food Forest will be a center for growth, connection, and celebration. The food forest concept is an ancient one, drawing on indigenous practices of caring for the land and people.

"The Festival Beach Food Forest will be a welcomed additional layer to sustainable food production in Austin."

Source: http://festivalbeach.org/about/

Quad City Food Forest, Davenport, Iowa

"The QC Food Forest mission: To design, grow, and maintain an edible landscape that will produce healthy food for the community, foster education, and support the environment.

"What are the goals of the QC Food Forest project?

- Promote quality of life in the Quad Cities
- Provide opportunities for the community to grow together
- Provide free food for the community
- Teach about native wild food and plants
- Provide a habitat for wildlife
- "Grow with us!"

Source: https://www.facebook.com/QC.FoodForest/

Rahma Edible Forest Snack Garden, Syracuse, New York

"Transform the cityscape from vacant lots to forested ecosystems, thereby contributing to biodiversity and providing green infrastructure that benefits the health of our watershed. Bring the health care mission of the Rahma Free Health Center outdoors, making free and fresh food available to everyone in the community."

Source: An informational sign at the Rahma Edible Forest Snack Garden

Bloomington Community Orchard, Bloomington, Indiana

"Bloomington Community Orchard is an organization devoted to growing fruit for the community and growing our orcharding skills through educational opportunities. The publicly owned orchard is maintained by volunteers, and the harvest is available to everyone in the community.

"The orchard contributes to Bloomington's food security, inspires joyful community engagement, and educates citizens while making sustainability delicious.

"**Mission:** Dream, build, and share an orchard community.

"**Vision:** To inspire communities to cultivate thriving systems of sharing and growing fruit.

"The Bloomington Community Orchard aspires to:

- Model how growing and sharing food supports food sovereignty, local resilience, and sustainability.
- Facilitate the sharing of knowledge and experience around perennial food growing practices.
- Support and advocate for the growing network of community orchards and individual growers.
- Cultivate a public commons that serves as an edible park and space for recreation, personal development, and joyful community engagement.

- Nurture a welcoming and diverse community, accessible to people of all identities, abilities, and experiences.
- Serve as a model of civic engagement that fosters leadership development, artistic expression, and continual learning.

"The Bloomington Community Orchard supports its organizational sustainability by:

- Maintaining accurate, ethical, and responsible financial practices.
- Conducting development programs that support its financial outcomes.
- Promoting programs and opportunities for involvement to the public.
- Developing leadership that is knowledgeable, critical thinking, and ethical.
- Operating as an all-volunteer nonprofit organization."

Source: http://www.bloomingtoncommunityorchard.org

Wetherby Edible Forest, Iowa City, Iowa

"The Wetherby Edible Forest inspires our community to gather and grow healthy food in ways that rehabilitate our local ecosystems while increasing equal access to food.

"Join us to improve public health by regenerating our public land into an edible forest ecosystem. We work to reduce agricultural climate impact, improve our local food security, provide educational opportunities, and celebrate growing food for the benefit of all species.

"All food is free for harvesting and eating. No synthetic pesticides or herbicides have been applied."

Source: http://www.backyardabundance.org/abundantlandscapes/wetherbyedibleforest.aspx

Tree Streets Food Forest, Johnson City, Tennessee

"A food forest is a type of community edible park that is grown based on permaculture principles. Permaculture is a holistic method of raising food that takes a whole systems approach to create self-sustaining, highly productive ecosystems on a given piece of land. This practice emphasizes edible perennial plants, building topsoil, and intensively managing all resources. Often, there is a lot of upfront work required to create a food forest, but the end result is a self-sustaining system that produces copious amounts of food, medicine, and other resources with little human intervention beyond harvesting."

Source: http://builditupetn.com/projects/food-forests/

Southern Heights Food Forest, Lincoln, Nebraska

"Our vision is to change the way we think about and value public space as a community. Creating a space where entire communities and families come together to interact with each other and build relationships. A space where children can play freely while learning about nature, where we teach each other about food, where people form multigenerational relationships—a catalyst for true community development. . . .

"The Southern Heights Food Forest includes a variety of elements that will change the way we eat, play, and interact as a community. The two-acre space will include Nebraska's first Food Forest, a Nature Explore outdoor classroom, urban agriculture plots, and an expansion of the existing community garden. By creating a dynamic space that is based around the value of food and the outdoors, we are creating a sanctuary for community development where children can play safely, where we learn about the value of food, and most important, where meaningful lifelong relationships are built within the Lincoln community."

Source: https://southernheightsff.org/

6th Ward Garden Park, Helena, Montana

"Our mission is to grow a garden park that connects people, food, and nature.

"Planning for the 6th Ward Garden Park began in 2013, and it continues to evolve into an inclusive, engaging space where the community can meet, learn, play, relax, grow, and enjoy food. We designed the park with nature in mind.

"As we grow, it is our hope that the innovative and ecological concepts used in the park's design will spread throughout the Prickly Pear Valley.

"The 6th Ward Garden Park will provide community garden plots for 6th Ward residents with a preference for low-income gardeners, with the goal of increasing access to healthy food and nutrition education."

Source: https://6thwardgardenpark.com/

Greenbelt Food Forest, Greenbelt, Maryland

"Greenbelt Food Forest has three interrelated goals:

1) To expand tree canopy cover by engaging local residents and students to attend community design sessions and plant native edible trees/shrubs, herbaceous perennials, and ground covers that feature permaculture design principles. The tree planting will reduce rapid runoff and erosion and provide food sources.
2) To assess and improve water quality of the streams around Springhill Lake using both expert consultation and students' observations and data collection. The expert will provide the initial assessment, work with students in collecting and analyzing the data, and provide a community report.
3) To facilitate multigenerational social and regenerative educational opportunities through collaboration with schools, after-school programs, the city government, and other volunteer-based groups. This collaborative approach supports the functions of each organization, linking our missions and program objectives.

Short-Term Goals

- Plant 27 trees, 70 shrubs, and 82 herbaceous perennials to create a food forest in an urban setting.
- Organize ecological hands-on educational workshops for adults and students.
- Enhance middle school science by linking their lab work to the outdoor classroom of the forest/rain gardens.
- Conduct stream baseline and ongoing water quality monitoring.

Long-Term Goals

- Establish an outdoor classroom to facilitate social and educational opportunities year-round.
- Increase the ecological understandings of residents and students.
- Harvest fruits, nuts, berries, medicinals, and produce from vines.
- Reduce air pollution, erosion, trash, and flooding.
- Improve water quality of the Springhill Lake Stream.
- Increase urban tree canopy and water retention through selection of plants.
- Enhance understanding of restoring Chesapeake landscapes and permaculture design."

Source: http://chears.org/foodforest/about/

Notes

Chapter 1: Community Food Forests on the Rise

1. Kyle H. Clark and Kimberly A. Nicholas, "Introducing Urban Food Forestry: A Multifunctional Approach to Increase Food Security and Provide Ecosystem Services," *Landscape Ecology* 28, no. 9 (2013): 1649–69; Melissa R. Poe, Rebecca J. McLain, Marla Emery, and Patrick T. Hurley, "Urban Forest Justice and the Rights to Wild Foods, Medicines, and Materials in the City," *Human Ecology* 41, no. 3 (2013): 409–22; Rebecca McLain, Melissa Poe, Patrick T. Hurley, Joyce Lecompte-Mastenbrook, and Marla R. Emery, "Producing Edible Landscapes in Seattle's Urban Forest," *Urban Forestry & Urban Greening* 11, no. 2 (2012): 187–94.
2. Geoff Andrews, *The Slow Food Story: Politics and Pleasure* (Montreal: McGill-Queen's University Press, 2008).
3. Roslynn Brain, "The Local Food Movement: Definitions, Benefits & Resources" (Logan, UT: Utah State University Extension Sustainability, 2012); Steve Martinez, Michael Hand, Michelle Da Pra, Susan Pollack, Luanne Lohr, Sarah Low, and Constance Newman, "Local Food Systems: Concepts, Impacts, and Issues," United States Department of Agriculture, 2010.
4. Barbara Kingsolver, Camille Kingsolver, and Steven L. Hopp, *Animal, Vegetable, Miracle: A Year of Food Life* (New York City: Harper Perennial, 2008).
5. Laura Lawson wrote a definitive history of community gardening, management strategies, and implications for their use on open urban space in *City Bountiful: A Century of Community Gardening* in America. This important book discusses the many values related to gardening throughout United States history. It draws attention to the relationship between the rise and decline of interest in gardening associated with times of social unrest and fiscal crisis. While gardening contributed to unifying the nation during times of war and economic downturn, it also associated the need for gardening with negative events. So, abandoning the need to garden symbolized the nation moving forward positively. Finally, Lawson's book provides a thorough explanation of why a perennial production mentality for public spaces did not previously exist.
6. Annie Lindstrom and Robyn Mello, "Occupy Vacant Lots," Blog Talk Radio, *Lifestyle* (blog), 2016.

Chapter 2: Systems Thinking for Community Food Forests

1. Donella H. Meadows, *Thinking in Systems: A Primer* (White River Junction, VT: Chelsea Green Publishing, 2008).
2. Nitrogen-fixing plants have extra nodules in their root systems, which house symbiotic bacteria that transform nitrogen found in the soil into a form usable by the plant. The plant brings the nitrogen to the surface through leaf growth, and when the leaves drop off the plant and decompose, they help to add nitrogen to the organic matter on the soil surface so nutrients are available to other plants as well.

3. Meadows, *Thinking in Systems*.
4. Cornelia B. Flora and Jan L. Flora, *Rural Communities: Legacy and Change* (New York: Routledge, 2007); Other contributors to the development of the framework are Mary Emery, PhD, South Dakota State University; Susan Fey and Corry Bregendahl, Iowa State University; Isabel Gutiérrez-Montes, Tropical Agricultural Research and Higher Education Center; and Edith Fernandez-Baca, United Nations Development Programme.
5. Ethan C. Roland, "Eight Forms of Capital," *Permaculture* 68 (January 2011): 58–61.
6. Toby Hemenway, *The Permaculture City: Regenerative Design for Urban, Suburban, and Town Resilience* (White River Junction, VT: Chelsea Green Publishing, 2015).
7. David Holmgren, *Permaculture: Principles and Pathways Beyond Sustainability* (Victoria, Australia: Holmgren Design Services, 2002).

Chapter 3: Capital Investments in Community Assets

1. Flora and Flora, *Rural Communities: Legacy and Change*.
2. Isabel Gutiérrez-Montes, Mary Emery, and Edith Fernandez-Baca, "The Sustainable Livelihoods Approach and the Community Capitals Framework: The Importance of System-Level Approaches to Community Change Efforts," *Community Development* 40, no. 2 (2009): 106–13.
3. Diego Thompson, "'Somos Del Campo': Latino and Latina Gardeners and Farmers in Two Rural Communities of Iowa—A Community Capitals Framework Approach," *Journal of Agriculture, Food Systems, and Community Development* 1, no. 3 (2016): 3–18.
4. Cornelia B. Flora, "Mobilizing Community Capitals to Support Biodiversity," in *The Importance of Biological Interactions in the Study of Biodiversity* (Rijeka, Croatia: INTECH Open Access Publisher, 2011).
5. Flora and Flora, *Rural Communities: Legacy and Change*.
6. Eva Gamarnikow, "Social Capital and Human Capital," in *Encyclopedia of Community: From the Village to the Virtual World*, edited by K. Christensen and D. Levinson, (London: SAGE Publications, 2003), 1286–91; Gary P. Green and Anna L. Haines, *Asset Building and Community Development* (Thousand Oaks, CA: SAGE Publications, 2015).
7. Green and Haines, *Asset Building*; Jules N. Pretty, *Agri-Culture: Reconnecting People, Land and Nature* (Sterling, VA: Routledge, 2002).
8. Yvonne Beran, Tina Burr, Charlie French, and Nancy Friese, "Community Gardening in New Hampshire from the Group Up," University of New Hampshire Cooperative Extension, 2012.
9. Green and Haines, *Asset Building*.
10. Scott R. Sweetland, "Human Capital Theory: Foundations of a Field of Inquiry," *Review of Educational Research* 66, no. 3 (1996): 341–59.
11. Thompson, "'Somos Del Campo.'"
12. Ibid.
13. Gamarnikow, "Social Capital and Human Capital."
14. Richard Florida, "The Rise of the Creative Class," *The Washington Monthly* 34, no. 5 (2002): 15–25.
15. Green and Haines, *Asset Building*.
16. Ibid.
17. Cornelia B. Flora, Mary Emery, Susan Fey, and Corry Bregendahl, "Community Capitals: A Tool for Evaluating Strategic Interventions and Projects" (Ames, IA: North Central Regional Center for Rural Development at Iowa State University, 2005).

Chapter 4: Planning Fundamentals

1. Green Deane, "New Jersey Tea," Eat the Weeds (website), 2011.
2. Mark Scialla, "Roger Williams Park Edible Forest Garden," senior honors project, University of Rhode Island, 2012.

3. Darrin Nordahl, *Public Produce: The New Urban Agriculture* (Washington, DC: Island Press, 2009).

Chapter 5: Planning to Create Change

1. *Whole Measures for Community Food Systems* is a very useful document authored by several leaders in community food initiatives and whole community design programs. Based on the Center for Whole Communities' 6th edition of *Whole Measures: Transforming Our Vision for Success* (2007), the publication extensively addresses the conditions, assumptions, needs, and creativity related to achieving success when outlining action frameworks for developing particular projects. It is worth the time for community food forest leaders and stakeholders to read. Moreover, it has an appendix with practical tools that will also help you get started working on clear outcomes in Whole Measures assessments in support of developing a theory of change. Plus, it is freely available online.
2. W. T. Oswald, *A Guide to Conducting Popular Education* (San Diego, CA: The Global Action Research Center, 2014).
3. K. Pezzoli, P. Watson, M. Rabinowitz Bussell, W. Oswalk, J. B. Hinds, R. A. Leiter, L. Levin, and M. Watson-Ching, *Urban Agriculture and Food Disparities: An Overview of Phase I Progress* (San Diego, CA: University of California San Diego Lab for Sustainability Science, 2015).

Chapter 6: Rooting in History

1. Rowan Jacobsen, "Bring Back the Strange Apples!" *The Boston Globe*, September 5, 2014.
2. "The Transition Town Movement," Transition United States, http://transitionus.org/transition-town-movement.
3. Steven Pavlos Holmes, *A Healing Landscape* (Lincoln, MA: Mass Audubon, 2016).
4. Cornelia B. Flora, J. L. Flora, and S. P. Gasteyer, *Rural Communities: Legacy and Change* (Boulder, CO: Westview Press, 2015).
5. Roy Rosenzweig and Elizabeth Blackmar, *The Park and the People: A History of Central Park*, (Ithaca, NY: Cornell University Press, 1992).
6. Liz Barnes, *Return to the Forest Where We Live*, (Baton Rouge, LA: Louisiana Public Broadcasting, 2009).
7. Christopher Alexander, *The Oregon Experiment*, vol. 3 (New York: Oxford University Press, 1975).

Chapter 7: The Dr. George Washington Carver Edible Park

1. "Foodtopia Asheville: A Creative Culinary Community," *Explore Asheville*, https://www.exploreasheville.com/foodtopia/.
2. Food Research and Action Center http://frac.org/.
3. Katy Nelson, "As More Struggle to Afford Healthy Food, Asheville-Buncombe Council Works to Collaborate, Make Change," *Carolina Public Press*, 2012.
4. *Food Hardship in America 2011: Data for the Nation, States, 100 MSAs, and Every Congressional District* (Washington, DC: Food Research and Action Center, 2012).
5. Resolution No. 13–17, January 22, 2013, https://drive.google.com/file/d/0B4BMRkxgEB0cVnFCU05pMmprdzQ/view
6. Betty Jamerson Reed, *School Segregation in Western North Carolina: A History, 1860s–1970s* (Jefferson, NC: McFarland & Company, 2011).
7. Shiloh Community Association http://www.shilohnc.org

Chapter 8: The Role of Agroecology

1. K. F. Wiersum, "Forest Gardens as an 'Intermediate' Land-Use System in the Nature–Culture Continuum: Characteristics and Future Potential," *Agroforestry Systems* 61–62, no. 1–3 (2004): 123–34.
2. Dave Jacke with Eric Toensmeier, *Edible Forest Gardens*, vol. 2, *Ecological Design and Practice for Temperate Climate Permaculture* (White River Junction, VT: Chelsea Green Publishing, 2005).

3. Dave Jacke with Eric Toensmeier, *Edible Forest Gardens*, vol. 1, *Ecological Vision and Theory for Temperate Climate Permaculture* (White River Junction, VT: Chelsea Green Publishing, 2005).
4. Charles Kingsley, *At Last: A Christmas in the West Indies* (London: Macmillan, 1872).
5. *How to Make a Forest Garden,* by Patrick Whitefield, is a practical and trustworthy first step focused on plant arrangement and stewardship. For a more thorough study of theory, design, practice, and species selection, the two-volume compilation titled *Edible Forest Gardens,* by Dave Jacke and Eric Toensmeier, is a highly respected resource. Several other hallmark works exist, but what is consistent across all these contributions is that food forest design varies in scale and scope. Equally clear in the books on food forest design is that the possibilities for polyculture production are substantial and the spirit of a plan is limited only by imagination.

Chapter 9: Allies in Creating and Managing Public Space

1. Diana Boros and J. Glass, eds., *Re-imagining Public Space: The Frankfurt School in the 21st Century* (New York City: Palgrave Macmillan, 2014).
2. Joan Iverson Nassauer, "Messy Ecosystems, Orderly Frames," *Landscape Journal* 14, no. 2 (1995): 161–70.
3. Joan Iverson Nassauer, ed., "Cultural Sustainability: Aligning Aesthetics and Ecology," in *Placing Nature: Culture And and Landscape Ecology* (Washington, DC: Island Press, 1997), 65-84.
4. Aldo Leopold, *A Sand County Almanac, and Sketches Here and There* (New York: Oxford University Press, 1989).
5. Christopher Alexander, *The Oregon Experiment,* vol. 3, Center for Environmental Structure (New York: Oxford University Press, 1975).
6. William H. Whyte, *The Social Life of Small Urban Spaces* (Washington, DC: The Conservation Foundation, 1980).
7. A helpful way to approach grant writing in regard to urban forestry and community food forests is to match objectives and language used in the National Urban and Community Forestry ten-year strategic plan found at http://urbanforestplan.org/key-issues-for-2016-2026/.
8. Kyle H. Clark and Kimberly A. Nicholas. "Introducing Urban Food Forestry: A Multifunctional Approach to Increase Food Security and Provide Ecosystem Services," *Landscape Ecology* 28, no. 9 (2013): 1649–69.
9. Darrin Nordahl, *Public Produce: The New Urban Agriculture* (Washington, DC: Island Press, 2009).
10. Melissa R. Poe, Rebecca J. McLain, Marla Emery, and Patrick Hurley, "Urban Forest Justice and the Rights to Wild Foods, Medicines, and Materials in the City," *Human Ecology* 41, no. 3 (2013): 409–422; Rebecca J. McLain et al., "Producing Edible Landscapes in Seattle's Urban Forest," *Urban Forestry and Urban Greening* 11, no. 2 (2012): 187–194.
11. Poe, et al., "Urban Forest Justice and the Rights to Wild Foods, Medicines, and Materials in the City"; Lisa Tyrväinen, Stephan Pauleit, Klaus Seeland, and Sjerp De Vries, "Benefits and Uses of Urban Forests and Trees," in *Urban Forests and Trees: A Reference Book* (Berlin: Springer-Verlag Berlin Heidelberg, 2005): 81–114.
12. *Urban and Agricultural Communities: Opportunities for Common Ground* (Ames, IA: Council for Agricultural Science, 2002).
13. Nordahl, *Public Produce.*
14. Dan Nosowitz, "From Farmland to Golf Course and Back Again," *Modern Farmer* (blog), 2015.
15. Taylor Danz, "A Plan to Turn Hiawatha Golf Course into Minneapolis' First Food Forest," *City Page*, March 15, 2017, http://www.citypages.com/restaurants/a-plan-to-turn-hiawatha-golf-course-into-minneapolis-first-food-forest/416059773.

16. Tesha M. Christensen, "Local Resident Envisions Food Forest at Hiawatha Golf Course," *Longfellow Nokomis Messenger*, 2015, sec. "Communication Information."
17. April Philips, *Designing Urban Agriculture: A Complete Guide to the Planning, Design, Construction, Maintenance and Management of Edible Landscapes* (Hoboken, NJ: John Wiley & Sons, 2013).
18. Mark Gorgolewski, June Komisar, and Joe Nasr, *Carrot City: Creating Places for Urban Agriculture* (New York: Monacelli Press, 2011).
19. Ibid.
20. David Hanson and Edwin Marty, *Breaking Through Concrete: Building an Urban Farm Revival* (Berkely: University of California Press, 2012).

Chapter 11: Reflecting on Community
1. Jacke and Toensmeier, *Edible Forest Gardens*.
2. Mark Horowitz, *The Dance of We: The Mindful Use of Love and Power in Human Systems* (Amherst, MA: Synthesis Center Press, 2014).
3. The nonprofit Urban Tilth partnered with the USDA Forest Service, West Contra Costa County Unified School District, and City of Richmond Parks Landscaping Division to acquire grants through the Forest Service's Pacific Southwest Research Station.

Chapter 12: Building Social Systems
1. Toby Hemenway, *The Permaculture City: Regenerative Design for Urban, Suburban, and Town Resilience* (White River Junction, VT: Chelsea Green Publishing, 2015).

Chapter 13: The Beacon Food Forest
1. Rebecca McLain, Melissa Poe, Patrick T. Hurley, Joyce Lecompte-Mastenbrook, and Marla R. Emery, "Producing Edible Landscapes in Seattle's Urban Forest," *Urban Forestry & Urban Greening* 11, no. 2 (2012): 187–94.
2. Garrett Hardin, "The Tragedy of the Commons," *Science* 162, no. 3859 (1968): 1243–48.
3. Elinor Ostrom, "Elinor Ostrom's 8 Principles for Managing a Commons," *Commons Magazine*, 2011.

Chapter 14: Collaborative Leadership
1. Peter G. Northouse, *Leadership: Theory and Practice* (Thousand Oaks, CA: SAGE Publications, 2018).
2. Robert K. Greenleaf, *Servant Leadership* (Mawah, NJ: Paulist Press, 1977).
3. Hermann Hesse, *The Journey to the East: A Novel* (New York: Noonday, 1956).
4. David Logan, John King, and Halee Fischer-Wright, *Tribal Leadership: Leveraging Natural Groups to Build a Thriving Organization* (New York: HarperCollins, 2008).
5. Petra Kuenkel, *The Art of Leading Collectively: Co-Creating a Sustainable, Socially Just Future* (White River Junction, VT: Chelsea Green Publishing, 2016).

Conclusion: Looking Back, Moving Forward
1. "About Habitat III," Habitat III: The United Nations Conference on Housing and Sustainable Urban Development, 2017.

Index

Note: Page numbers in italics refer to figures and photos. Page numbers followed by *t* refer to tables. Page numbers preceded by *ci* refer to images in the color insert.

ABFPC (Asheville Buncombe Food Policy Council), 96
action taker role, 175
active listening, 164
activism
   in community organizing, 162–63
   environmental awakening, 94
adaptability
   feedback loops and, 28–29
   importance of, 10, 59, 220
   role in permaculture and agroforestry, 111, 115, 117–18
   systems thinking and, 24, 59
adaptive management, 59–60
advisors, tips for finding, 57
advocate role, 175, 188, 223
aesthetics
   food forest benefits, *3*, 7–8, *119*, 166
   landscape as language, 123–24
   urban design contributions, 128–29
   See also artistic installations
Afro-Caribbean culture, food forest use, 114
agroecology, 111–120
   agroforestry contributions, 111–15, *113*, 116–18
   combining approaches, 116–18
   defined, 21
   permaculture contributions, 111–12, 115–18
   principles of, 111–12
   Wetherby Edible Forest case study, 116, 118–120, *119*
agroforestry
   defined, 21–22
   forest garden definition, 114
   role in community food forests, 111–15, *113*, 116–18
   Wetherby Edible Forest case study, 119–120
Alexander, Christopher, 128
alley cropping, 112
Alliance for Community Trees, 211, 213, 214
allies in public space management, 121–141
   benefits of professional allies, 121–22
   conversation with professional allies, 132
   Cultivate Kansas City case study, 138–141, *139*
   landscape architecture contributions, 124–28, *125–27*
   lessons learned, 223
   overview, 121
   urban agriculture contributions, 133–35, *134*
   urban design contributions, 128–131, *131*
   urban forestry contributions, 132–33
   urban gardening contributions, 135–36
   urban wildlife specialist contributions, 136–38, *137*
alternative economies, 218–19
American Community Gardening Association, 21
Animal, Vegetable, Miracle (Kingsolver), 23
apple cultivation, historical context, 83
apprenticeship programs
   Cultivate Kansas City, 138
   Richmond Edible Forest Project, 166–67
Arbor Day, 90
Arbor Day Foundation, 213, 226
archiving project information. *See* documentation
Arroyo Food Co-Op, 130
artistic installations
   Basalt Food Park, 150

INDEX

artistic installations (*continued*)
  Beacon Food Forest, *ci6*
  Bloomington Community Orchard, *210*, 213
  Freedom Square Food Forest, *ci5*, 65, *65*
  fruit trees art installation, *20*
  Swale floating food forest, 217
  urban design contributions, 128
The Art of Leading Collectively (Kuenkel), 203
Asheville, North Carolina
  about, 96
  cultural history of, 101
  first forestry school, 92
  See also Dr. George Washington Carver Edible Park
Asheville Buncombe Food Policy Council (ABFPC), 96
Asheville GreenWorks, 106
asset-based development, 35, 45–46
asset mapping
  landscape architecture use of, *125*
  Ocean View Growing Grounds case study, 76, 79–80
  purpose of, 46
assigned vs. emergent leadership, 197–99
assumptions
  checking with community input, 76
  power dynamics and, 165
  in theory of change, 71, 72, 73–74
Atlanta, Georgia, Browns Mill Food Forest, 225
authentic and servant leadership, 199–200

automobile, impact on society, 93, 94

backward mapping, 73
Backyard Abundance, 119–120
Basalt, Colorado, about, 143–44
Basalt Food Park
  building support and partnerships, 143, 144–46, *145*
  continuation and future visions, 149–150
  fast facts, 143
  gathering spaces, *ci15*
  idea generation, 144
  importance of flexibility, 220
  installation and community engagement, 146–49, *148*
  signage and fencing, *ci8*
  student field trip to, *ci14*
  urban wildlife concerns, 138, 146, 172
Basalt Public Library, 145
Basalt Seed Library Project, 144–45, *145*, 149
Beacon Food Forest, 183–196
  beehives, 138
  bee sculpture, *ci6*
  Bloomington Community Orchard relationship, 213
  building support and partnerships, 186–191, *187*
  continuation and future visions, 195–96, *196*
  cross-zone relationships, 179
  emergent leadership in, 198
  fast facts, 183
  gathering and event spaces, *ci16*
  idea generation, 184–86
  installation and community engagement, 191–95, *194*

land acquisition, 188–191, 220
  media attention, 18, 181, 183, 191, 192, 193, 220–21
  mission statement, 231
  mushroom production, *ci2*
  organizing people, 163
  participant motivations, 164–65
  polyculture plantings, *ci3*
  signs, *ci8*, *ci9*
  social media use, 57
  volunteer management, 222
  work day teamwork, *154*
beehives and bee boxes, *63*, 138
beneficial organisms, *137*
BFFC (Boston Food Forest Coalition), 83–86
Big Four Questions, 203
Biltmore Estate, 92, 101
Blackhawk Food Forest, *169*
Bloomington, Indiana, about, 205–6
Bloomington Community Orchard, 205–16
  about Bloomington, 205–6
  bonding capital, 153–54
  building support and partnerships, 48, 207–8
  celebrations, 41, 213–14, 223
  continuation and future visions, 212–16
  empowerment of volunteers, 165
  fast facts, 205
  gathering and event spaces, *ci15*
  governance structure, 221–22
  idea generation, 206–7, *207*
  initial emphasis on fruit trees, 221
  installation and community engagement, 208–12, *210*, *212*

INDEX

layers in, *ci2*
leadership examples, 198, 201
mission statement and vision, 232–33
variety in experiential conditions, *ci12*
virtual network usefulness, 181
Bloomington Food Policy Council, 214
blueberries, *104*
boards of directors or trustees
   Bloomington Community Orchard, 209–10, 212–13
   Bountiful City Project, 102
bonding capital, 36, 153–55, *154*
Boston Food Forest, 83–86, *86*
Boston Food Forest Coalition (BFFC), 83–86
Boston Nature Center, 84, 85–86
Boston Tree Party, 21
Bountiful Cities, 101, 105, 106
Bountiful City Project, 97, 98
   See also Dr. George Washington Carver Edible Park
Breaking Through Concrete (Hanson and Marty), 134
bridging capital, 36, 153–55, *154*, 157
Brier, Amy, 210
Brown, Jonathan
   building support and partnerships, 99–102
   idea generation, 96–98
Browns Mill Food Forest, 225
Brt, Amy Rose, 173–74
built capital, 43–44
Buncombe County Tourism Development Authority, 96
Buncombe Fruit and Nut Club, 105–6

California
   agricultural incentive zones, 46
   18th and Rhode Island Permaculture Garden, *ci13*, 46, 48, 172, 231
   Fallen Fruit, *20*, 21, 68
   Mesa Harmony Garden, 10–11, *ci16*, 57, *171*, 178, 203–4
   Ocean View Growing Grounds, 75–80, *78*
   Richmond Edible Forest Project, 48–49, 165–67, *166*
   Throop Unitarian Universalist Church food forest, *ci3*, 130–31
   Village Homes, 20
California Rare Fruit Growers, 203–4
canopy layers
   disturbances to, 223
   Dr. George Washington Carver Edible Park, 221
   Fargo Forest Garden, *ci4*
   filling in, 31, 106, 114
   management of, *105*, 106
capital investments in community assets, 35–49
   asset-based development, 45–46
   Basalt Food Park, 146
   built capital overview, 43–44
   collaboration in, 163–65
   cultural capital overview, 40–42
   examples of, 46–49, *47*
   financial capital overview, 44–45
   human capital overview, 37–38
   multiplication of gains, 227–28
   natural capital overview, 38–40
   need for community support, 158–59
   political capital overview, 42–43
   social capital overview, 35–36
   See also specific types of capital
capitals, stocks and flows of. *See* stocks and flows of community capitals
Caribbean culture, food forest use, 114
Carrot City: Creating Places for Urban Agriculture (Gorgolewski et al.), 133
Carver, George Washington, 97
Carver Edible Park. *See* Dr. George Washington Carver Edible Park
celebrations
   Bloomington Community Orchard, 213–14, 223
   built capital considerations, 44
   cultural capital considerations, 41–42, 157
   lessons learned, 223
   See also work parties and volunteer days
Center for Theory of Change, 73, 74
Center for Whole Communities, 75, 237n1
Central Park (New York City), 89–90
Cetera, Frank, 48
champion role, 175, 188, 223
change
   defining the vision for, 77–78
   Dr. George Washington Carver Edible Park case study, 95
   theory of, 71–75, *72*
charrettes
   Dr. George Washington Carver Edible Park, 102–4

INDEX

charrettes (*continued*)
feedback loops for, 28
landscape architecture contributions, 125–26
Ocean View Growing Grounds, 79
Choice Neighborhoods Grant program, 128
City Bountiful: A Century of Community Gardening (Lawson), 235n5
City Seeds
building support and partnerships, 98–102
formation of, 98
installation and community engagement, 102–5
name change, 74, 98
See also Dr. George Washington Carver Edible Park
civic agriculture movements, 17–18
Clark-Cooper Community Garden, 84–85
closure phase, 58–59
collaborative community leadership, 201–4
See also leadership
Collective Leadership Compass, 203
Colorado, Basalt Food Park. *See* Basalt Food Park
Colorado Division of Parks and Wildlife, 146, 147
Colorado Rocky Mountain Permaculture Institute (CRMPI), 144, 145
the commons, 192
Common Thread garden (Beacon Food Forest), 193

communication plans
community connection concerns, 68–69
establishment phase, 58
levels of, 57
maintenance phase, 66, 67
planning phase, 56
See also online presence
community, 153–167
bonding and bridging capital, 36, 153–55, *154*, 157
Fargo Forest Garden case study, 159–162, *160*
investing skills, 163–65
legacy of, 226–28
meaning of, 155–56
meeting needs of, 157–58
organizing people, 162–63
Richmond Edible Forest Project case study, 165–67, *166*
support from, 158–59
virtual, 181
Community Capitals Framework
Latino community gardener sociology study, 37
systems thinking for, 33*t*, 34
See also capital investments in community assets
community connection
agroecology benefits, 112
Basalt Food Park, 147
Beacon Food Forest, 189–190, 191–95, *194*
Bloomington Community Orchard, 208–12, *210*, *212*, 213, 214
challenges of, 122
community garden contributions, 135–36

Dr. George Washington Carver Edible Park, 102–5, 107
information kiosks and bulletin boards, *ci11*, 67, 69
knowing your community, 147
landscape architecture contributions, 126–27
long-term success dependence on, 11
need for interproject connectivity, 122, 226
outreach to local residents, 42–43, 76–77, 141, 147, 155–56
role in planning, 68–69
urban design contributions, 129–130
community context, during initiation phase, 55
Community First! Village (Austin, Texas), 20–21
community food forests
defined, 4–6, *5*
design of, 7–10
increasing popularity of, 17–24
interaction among, 122, 226
purpose of, 6–7
community gardens
Asheville, North Carolina, 97, 98
Bloomington, Indiana, 208
Clark-Cooper Community Garden, 84–85
contributions to public spaces, 135–36
early twentieth century, 91, 92
Goodwill Industries of the Valley community garden, 127–28
P-Patch community garden program, 135, 188–191

• 244 •

INDEX

as resource for finding advisors, 57
Shiloh community, 101
Urban Gardening Program, 21
Community Leaders Fellowship, 84
community organizing, 162–63
  See also organizing people
compassionate communication, 164
complex systems, 27–28
  See also systems thinking
conceptual skills, 200, 201
conflict management
  in core groups, 176–77
  resources for, 164
connections, between system elements, 25, 26–27
  See also community connection
Conservation Corps, 190
context, during initiation phase, 55
core groups
  Basalt Food Park, 143, 145, 146, 148–49
  Beacon Food Forest, 186
  Bloomington Community Orchard, 153–54, 209–12
  bonding and bridging capital considerations, 153–55
  Boston Food Forest, 83–84
  collaborative community leadership in, 202
  communication for, 57–58
  Fargo Forest Garden, 159–161
  Freedom Square Food Forest, 63
  key roles in, 175–76
  organizing people and, 162
  scope of project and, 56
  site installation phase, *169*
  as Zone 0, 174–77

See also leadership
Countryman, Amy, 206–10
Cradle of Forestry, 92
Cramer, Jacqueline, 184, 185, 186, 187
critical discourse analysis, 156
critical reflection. *See* reflection
CRMPI (Colorado Rocky Mountain Permaculture Institute), 144, 145
Cultivate Kansas City, 138–141, *139*
cultural capital
  Basalt Food Park, 147
  connection to bonding capital, 155
  Dr. George Washington Carver Edible Park, 101
  overview, 40–42
  role in collaborative community leadership, 202–4
cultural diversity, meeting needs of, 157–58

The Dance of We (Horowitz), 164
Davis, Larry, 139
Dean, Sherman, 166–67
debt capital, 44–45
decision making
  collaborative community leadership and, 202
  direct participation, 43
  scope of project and, 56
demonstration projects, cultural capital considerations, 40–41
DePave (organization), 159, 161, 162, 178
depaving, 159–160
Dermitzel, Daniel, 138, 139–140
designer role, 175, 176

Designing Urban Agriculture (Philips), 133
design process
  Beacon Food Forest case study, 185–86, 189–191
  charrettes, 28, 79, 102–4, 125–26
  Cultivate Kansas City case study, 138–141, *139*
  Fargo Forest Garden case study, 159–162, *160*
  Goodwill Industries of the Valley community garden case study, 127–28
  landscape architecture contributions, 124–28
  natural systems influence on, 27
  social design considerations, 7–10, 170
  Throop Unitarian Universalist Church food forest case study, 130–31, *131*
  urban agriculture contributions, 133–35, *134*
  urban design contributions, 128–130
  urban forestry contributions, 132–33
  urban gardening contributions, 135–36
  urban wildlife specialist contributions, 136–38
  Wetherby Edible Forest case study, 56, 116, 118–120, *119*, 233
  See also agroecology; systems thinking
Detroit, Michigan, community gardens, 91, 92
dialogue with agencies and the public, lessons learned, 220–21

INDEX

Diamond, Michael, 123
DiNardo, Lisa, 145, 146–47, 149
direct participation, 43
disturbances, effects of, 223–24
diversity
    in core groups, 176
    cultural capital considerations, 40
    meeting community needs, 157–58, 227
documentation
    Dr. George Washington Carver Edible Park, 99
    importance of, 58, 99
    list of assumptions and activities, 74
Dr. George Washington Carver Edible Park, 95–108
    building support and partnerships, 98–102, *100*
    continuation and future visions, 105–8
    fast facts, 95
    founding organization name change, 74
    idea generation, 96–98, *97*
    installation and community engagement, 102–5, *104*, *105*
    lack of media attention, 191
    layers in, *ci12*, 221
    mission statement, 231
    newsletters, 58
    scope of change over time, 223
dry regions, food forest layers for, 31
Dula, Jimmy, 148, 149, 150

Eames, Charles, 26
Eames, Ray, 26
Earth Fare, 104

ecological functions, defined, 39
ecology, as term, 90
eco-mimicry, 111
economic downturn (2008-2009), 23
economies, alternative, 218–19
ecosystems
    awareness of, 90
    community food forests as, 27–28, 29–31, *30*
Eddins, Darcel, 105
Edible Agroforestry Design Templates manual (Backyard Abundance), 119
Edible Entropy, 74, 98
    See also City Seeds
Edible Forest Gardens (Jacke and Toensmeier), 160, 238n5
edible landscaping
    examples of, 20
    landscape as language, 123–24
    as term, 158–59
education
    Basalt Food Park programs, 149–150
    Beacon Food Forest programs, 194, 222
    Bloomington Community Orchard programs, 213, 214, 216
    Cultivate Kansas City programs, 138
    Dr. George Washington Carver Edible Park programs, 102, 103
    from food forests, 6–7
    human capital considerations, 37–38
    Ocean View Growing Grounds programs, 78–79, *78*
    popular, 77

    See also youth involvement
Edy's Ice Cream, 208–9, 211
18th and Rhode Island Permaculture Garden
    building support and partnerships, 46, 48
    land acquisition, 172
    mission statement, 231
    terraces and contoured pathways, *ci13*
electricity, arrival of, 91
elements, of systems, 25
emergence, as concept, 26
emergent vs. assigned leadership, 197–99
empowerment benefits of community food forests, 19, 165, 222
environmental awareness, growth of, 90, 94
environmental services
    defined, 39
    food forest benefits, 7–8
equilibrium considerations, 24
equity capital, 44–45
establishment phase
    Basalt Food Park, 146–49, *148*
    Beacon Food Forest, 191–95, *194*
    Bloomington Community Orchard, 208–12, *210*, *212*
    decision making on layers, 31
    Dr. George Washington Carver Edible Park, 102–5, *104*, *105*
    Fargo Forest Garden, 116
    grant funding for, 44, 45
    inclusion of quick-fruiting species, 68
    Ocean View Growing Grounds case study, 79

INDEX

overview, 57–58
   social capital considerations, 36
ethnobotany, use in agroecology, 111
European colonization, 87–88
event spaces. *See* gathering and event spaces
Evergreen Village (Bloomington, Indiana), 215
execution phase. *See* establishment phase

facilitator role, 175, 176–77
Fallen Fruit, *20*, 21, 68
FAO (Food and Agriculture Organization), 224–25
Fargo Forest Garden
   adjacent café, *ci7*
   canopy layer, *ci4*
   community support for, 159–162, *160*
   establishment phase, 116
   soil mounds, *ci5*
feasibility, initiation phase considerations, 56
feedback loops
   adaptive management and, 59
   initial design, 125
   overview, 28–29
fencing
   Basalt Food Park, *ci8*, 146–47, 149, 172
   Bloomington Community Orchard, 208, 209
   fruit trees art installation, *20*
Festival Beach Food Forest, 232
financial capital
   collaboration among stakeholders and, 163–64
   investments in, 44–45

financial planning, 57
Fischer, Burney, 206, 207, 208
flexibility, need for, 28–29, 118, 220
   See also adaptability
Flora, Cornelia, 34, 35
Flora, Jan, 34, 35
flows. *See* stocks and flows of community capitals
food, as element of community food forests, 4–6, *5*
Food and Agriculture Organization (FAO), 224–25
Food Forest International Research Network, 224
food forests, as term, 114
   See also community food forests
food justice, 17–18
food literacy, 6–7, 38
Food Research and Action Center, 96
Foodtopian Society, 96
foraging and gleaning, urban, 133
Forested, LLC, 45
forest farming, 112
Forest Gardening (Hart), 22, 92
forest gardens
   definitions of, 114
   as element of agroforestry, 112–14
forestry schools, 92
forests
   characteristics of, 8
   as element of community food forests, 4–6, *5*
formal connections among forests, need for, 226
Free, Lady Gloria Howard, 101
Freedom Square Food Forest artistic installation, *ci5*

community capitals in, 48
   as nested project, 62–63, 65–66, *65*
   systems map, *47*
Friends of the Beacon Food Forest, 187, 191, 192
Fruit Tree Planting Foundation (FTPF), 208–9
Fukuoka, Masanobu, 92
function, of systems, 25, 27–28
funding
   agency relationships, 179
   Basalt Food Park, 147
   benefits of partnerships, 211–12
   Bloomington Community Orchard, 208–9, 213
   Dr. George Washington Carver Edible Park, 98
   Fargo Forest Garden, 161
   financial capital considerations, 44
   increasing opportunities for, 225–26
   need for long-term support, 69–70
   See also grant funding
fundraiser role, 175

Garden City movement, 92
gathering and event spaces
   Basalt Food Park, *ci15*, 150, 193
   Beacon Food Forest, *ci16*, *194*
   Bloomington Community Orchard, *ci15*
   as built capital, 44
   Dr. George Washington Carver Edible Park, 101
   maintenance of, *66*
   Mercy Edible Park, *ci4*

INDEX

gathering (*continued*)
  Mesa Harmony Garden, *ci16*
  Ocean View Growing Grounds, 79
  permaculture zone concept and, 116
Gauthier, Leah, 215
Gay, Ross, 201
Genesis Gardens (Austin, Texas), 20–21
George Washington Carver Edible Park. *See* Dr. George Washington Carver Edible Park
Georgia
  Browns Mill Food Forest, 225
  Serenbe Community, 20
Georgiou, Bob, 76, 78
Giangrande, Naresh, 22
Gibbs Road Farm, 138, 140
gleaning and foraging, urban, 133
Global Action Research Center, 77, 78, 79
goal-setting, in permaculture design methodology, 124
Goldsmith, Angela, 159–160
Goodwill Industries of the Valley community garden, 127–28
governance structure
  Collective Leadership Compass assistance with, 203
  core group roles, 56, 174–77
  lessons learned, 198, 221–22
  scope of project and, 56
  See also leadership
grant funding
  Beacon Food Forest, 188–89, 190, 193, 195
  Bloomington Community Orchard, 208–9, 211, 213, 214, 221–22
  for establishment phase, 44, 45
  need for follow-on support, 69–70
  suggestions for, 238n7
Great Recession (2008-2009), 23
Greeley Food Forest, 162
Green Ambassadors program, 218
Greenbelt Food Forest, *113*, 234
Greenleaf, Robert K., 199
Green Revolution, 93
GreenWorks, 106
GROW Northwest, 188, 189
Guidelines on Urban and Peri-urban Forestry (UN FAO), 224–25

Habitat (United Nations conference), 224–25
Habitat for Humanity, 215
Haeckel, Ernst, 90
Hale, Stephen, 218–19
Hale-Y Community Garden Food Forest
  community connection, 136
  cross-zone relationships, 179
  diversity of participants, 41
  signs, *ci11*
Hanson, David, 134
Happy Farms, 140
Hardin, Garrett, 192
Harrison Design, 189
Hart, Robert A., 22, 92
harvesting, pitfalls to avoid, 66–67, *66*
Hazelwood Food Forest
  capital investments, 64
  monarch butterfly habitat, 137–38, *137*
  replacement with plant nursery, 9, *9*, 10
  site rehabilitation, 39

  visit to, 1–4, *2*, *3*
A Healing Landscape (Holmes), 85
Helena Community Gardens, 177
Hemenway, Toby, 34, 172
herb spirals, *62*, *86*
heritage plants, 215
Herlihy, Glenn, 184–86, 187, 191, 193–94, 195, 222
Hesse, Hermann, 199
historical considerations for planning, 81–94
  1600s and 1700s, 87–88
  1700s and 1800s, 88–91
  1900s, 92–94
  agroforestry use, 112
  Boston Food Forest case study, 83–86, *86*
  Dr. George Washington Carver Edible Park case study, 99–101
  key questions, 81–82
  perspective, 82–83
  reflection and insights, 82
holistic education approach, 38
Holmes, Steven Pavlos, 85
Holmgren, David, 22, 34, 115
Hoosier Hills Food Bank, 211, 213
Hopkins, Rob, 22
Horowitz, Mark, 164, 165
Howard, Ebenezer, 92
How to Make a Forest Garden (Whitefield), 238n5
hügelkultur, *62*, *86*
human capital
  of individuals, 173–74
  principles of, 37–38
Huss, Lee, 207–8

idea generation
  Basalt Food Park, 144

INDEX

Beacon Food Forest, 184–86
Bloomington Community Orchard, 206–7, *207*
Dr. George Washington Carver Edible Park, 96–98, *97*
inclusiveness
  bridging capital and, 153, 154–55
  in core groups, 176
  importance to community building, 171, 227
independent vs. nested projects, 60–63
Indiana, Bloomington Community Orchard. *See* Bloomington Community Orchard
Indiana University, 48, 206, 214, 215, 216
individual, as Zone 00, 173–74
industrial agriculture
  historical considerations, 88–89
  nature vs., 90–91
industrialization, 90
information kiosks and bulletin boards, *ci11*, 67, 69
Inhabit Earth, 225–26
initiation phase, 54–56
Inland NW Food Forest Council Partners, 184
inputs, short- and long-term
  Mercy Edible Park, *133*
  Ocean View Growing Grounds case study, 78–79
installation phase. *See* establishment phase
instrumental political capital, 42
intentional communities, community food forest roots in, 20–21

intentionality, in roles, 164, 176
International Society of Arboriculture, 132
international support for urban food forestry, 224–25
interpersonal skills, 200, 201
Iowa
  Blackhawk Food Forest, *169*
  Quad City Food Forest, 69, 232
  Wetherby Edible Forest, 56, 116, 118–120, *119*, 233
Iowa City Parks and Recreation, 119–120
Iowa Department of Agriculture, 119
Irby, Jim, 222

Jacke, Dave, 160, 238n5
Jacobsen, Rowan, 83
Jamaica Plain neighborhood, 83–84
Jamaica Plain New Economy Transition (JP NET), 84
Jefferson Park, 185
Jefferson Park Alliance, 185, 188
Jefferson Park Jubilee, 186, 192
The Journey in the East (Hesse), 199
JP NET (Jamaica Plain New Economy Transition), 84

Kansas, Cultivate Kansas City project, 138–141, *139*
Kelly, Katherine, 138
Kids in the Woods grant program, 165
King, Franklin Hiram, 92
Kingsley, Charles, 114
Kingsolver, Barbara, 23

Kopf, Al, 99
Kriegman, Orion, 84
Kuenkel, Petra, 203

labeling, of plants, *ci10*, 43
landowners, finding, 54, 172–73
landscape architects, as public space allies, 124–28, *125–27*
landscape as language, 123–24
land selection. *See* site selection
Land Stewardship Specialty Crop Block Grant Program (Iowa), 119
large-scale agriculture
  historical considerations, 88–89
  nature vs., 90–91
Latino community gardener sociology study, 37
laws and regulations
  aid from professional allies in meeting, 124, 223
  Basalt Food Park challenges, 146
  Beacon Food Forest challenges, 188, 191
  FAO resource on, 224
  political capital and, 42–43
  public policy concerns, 67–68
Lawson, Laura, 235n5
layers, in food forests
  Bloomington Community Orchard, *ci2*
  Cultivate Kansas City, 140
  Dr. George Washington Carver Edible Park, *ci12*, 221
  ecosystems view of, 29
  species for, 31*t*
  tropical home gardens, 112–13
  typical numbers at beginning of projects, 31

INDEX

layers, in permaculture, 115
leadership, 197–204
    assigned vs. emergent, 197–99
    collaborative, 201–4
    role in building social systems, 170
    servant and authentic, 199–200
    skills and traits of, 199, 200–201
    turnover in, 55
    See also core groups
Learning/Action Center (Ocean View Growing Grounds), 79–80
Lefko, Samantha
    building support and partnerships, 99–102
    idea generation, 96–98
Leopold, Aldo, 124
lessons learned, 219–224
leverage points, in change, 29
liability insurance organizations, 179
life-affirming systems, 164, 165
lighting considerations, 107–8
local food
    community food forest links, 17–18
    historical contexts, 82–83

magician role, 175
maintenance. *See* monitoring and maintenance phase
managing vs. leading, 198–99
Mann, Stephen, 141
Marshall strawberry, 215
Marty, Edward, 134
Maryland, Greenbelt Food Forest, *113*, 234
Massachusett people, 85
Massachusetts

Boston Food Forest, 83–86, *86*
Boston Tree Party, 21
Massachusetts Audubon Society, 84, 85
Master Gardeners
    Mesa Harmony Garden involvement, 203–4
    as potential Zone 1 volunteers, 178
    Roger Williams Park Edible Forest Garden involvement, 60–62
master plans, 126
Mattingly, Mary, 217
Meadow, Donella
    stocks and flows of community capitals, 32
    systems thinking, 25
meaningfulness
    cultural capital considerations, 40
    food forest benefits, 18, 227
measurement of outcomes, 74–75, 102
media relations
    Beacon Food Forest, 18, 181, 183, 191, 192, 193, 220–21
    lessons learned, 220–21
    as Zone 4, 179, 181
Meierding, Allison, 84
Mello, Robyn, 46
Mercy Edible Park, *ci4*, 46, *133*
Mesa Harmony Garden
    community connection, 10–11
    gathering spaces, *ci16*
    intergenerational learning and sharing, 178
    site maintenance, 203–4
    weekly emails, 57
    work parties, *171*

metrics for evaluation. *See* measurement of outcomes
Meyer, Fred, 119
mission
    Bloomington Community Orchard, 214
    examples of, 231–34
    initiation phase considerations, 55
Mollison, Bill, 22, 115
monarch butterfly habitat, 137–38, *137*
monitoring and maintenance phase
    adaptive management and, 59
    Beacon Food Forest, 195–96
    Bloomington Community Orchard, 212–16
    canopy layer thinning, 221
    Dr. George Washington Carver Edible Park, 107, 221
    Mesa Harmony Garden, 203–4
    need for community support, 158–59
    one-time vs. ongoing efforts, 159
    overview, 58
    pitfalls to avoid, 66–67, *66*
    sustaining momentum, 159
Montana, 6th Ward Garden Park, 173, 177, 234
Montford neighborhood (Asheville, North Carolina), 104–5
Mother Hubbard's Cupboard, 211, 213
motivator role, 175–76
multistory cropping, 113–14
mushroom production, *ci2*
music garden, at Basalt Food Park, 150

INDEX

names, importance of, 98
Nassauer, Joan Iverson, 123–24
national collaboration among community food forests, 122, 226
Native Americans
   1600s and 1700s, 87–88
   agroforestry use, 112
natural capital
   collaboration among stakeholders and, 163–64
   Native American vs. European colonist views, 87
   overview, 38–40
natural systems, 27
Nebraska, Southern Heights Food Forest, 173–74, 233–34
Neighborhood Food Network, 79
Neighborhood Matching Fund, 189
Neighborhood Orchards Program, 215
nested systems
   elements of, 29–31, *30*
   independent vs. nested projects, 60–63
   Powers of Ten example, 26
   urban forestry contributions, 133
networker role, 175
New England Resilience and Transition Network, 84
New York City
   community gardens, 91
   green spaces, 89
   Rahma Edible Forest Snack Garden, *ci1*
   Swale floating food forest, 217
New York State
   Freedom Square Food Forest, *ci5*, *47*, 48, 62–63, 65–66, *65*

Rahma Edible Forest Snack Garden, 48, 232
1900s US food history, 92–94
nitrogen-fixing plants, 235n2
nonedible species, roles of, *ci1*
nonnative species, *ci3*
nonviolent communication, 164
Nordahl, Darrin, 67, 133
North Carolina
   about Asheville, 96, 101
   Shiloh community, 101
   See also Dr. George Washington Carver Edible Park
Northouse, Peter, 197
note taker role, 175, 176

Obama, Michelle, 18
Occupy Vacant Lots, 23
Ocean View Growing Grounds, 75–80, *78*
Olmsted, Frederick Law, 83–84, 89, 91, 185
One Hundred Strong, 76
one-time events or engagements, 159, 179, 181
online presence
   benefits of, 181, 191
   Bloomington Community Orchard, 209, 213
   for project documentation, 99
   social media usefulness, 42, 99, 181
   virtual communities, 181
Oregon
   Fargo Forest Garden, *ci4*, *ci5*, *ci7*, 116, 159–162, *160*
   Greeley Food Forest, 162
   Urban Farm Collective, 161–62
The Oregon Experiment (Alexander), 128

Oregon Sustainable Agriculture Land Trust, 162
organic agriculture movement, 93
organizational structure. *See* governance structure
organizing a community, 163
organizing people
   community needs, 162–63
   importance of, 10
   social media usefulness, 181
Osentowski, Jerome, 144
Ostrom, Elinor, 192
Oswald, Bill, 77
outcomes
   balancing for community needs, 134
   benefits of professional allies, 122
   measurement of, 74–75, 102
   Ocean View Growing Grounds case study, 79–80
   in theory of change, 71, *72*

parks, public-private partnerships with, 31–32
partnerships
   Basalt Food Park, 144–46, 149
   Beacon Food Forest, 186–87, 188–191
   benefits of, 143, 189
   Bloomington Community Orchard, 207–8, 211–13, 214
   Dr. George Washington Carver Edible Park, 98–102, *100*
   engaging in dialogue, 220–21
   landownership considerations, 172–73
party planner role, 175, 176
pathways
   design considerations, 116
   Wetherby Edible Forest, *119*

INDEX

pavement, removing, 159–160
Pearson Garden, 105
Pell, Jenny, 185, 189
Pennsylvania
   community gardens, 91
   Hazelwood Food Forest, 1–4, *2, 3*, 9, *9*, 10, 39, 64, 137–38, *137*
   Mercy Edible Park, *ci4*, 46, *133*
   Philadelphia Orchard Project, 21
people skills, 200, 201
perceived rules, restrictions from, 42–43
perennial food production, new models for, 21–22
peripheral agencies, 179
permaculture
   application to social systems, 171–72
   Boston Food Forest case study, 84–85, *86*
   design methodology, 124–25
   ethics and design principles, 33*t*, 34
   forest garden definition, 114
   overlap with landscape architecture, 124
   rise of, 22
   role in community food forests, 111–12, 115–18
   Wetherby Edible Forest case study, 119–120
The Permaculture City (Hemenway), 34, 172
perseverance, importance of, 220
Peterson, Jessica, 173, 177
Peterson Garden Project, ix–x
Pezzoli, Janice, 76
Pezzoli, Keith, 75–77
Philadelphia, Pennsylvania
   community gardens, 91
   Mercy Edible Park, *ci4*, 46, *133*
   Occupy Vacant Lots, 23
Philadelphia Orchard Project, 21
Philips, April, 133
physical (built) capital, 43–44
Pisgah National Forest, 92
pitfalls to avoid, 63–70
   community connection concerns, 68–69
   funding concerns, 69–70
   maintenance problems, 66–67, *66*
   overview, 63–64
   public policy concerns, 67–68
   safety concerns, 64–66, *65*
planning, 53–80
   adaptive management, 59–60
   bonding capital during, 153–54, 155
   checking assumptions with community input, 76
   defining the vision for change, 77–78
   human capital considerations, 37–38
   importance of, 10
   independent vs. nested projects, 60–63
   need for collaboration in, 170
   Ocean View Growing Grounds case study, 75–80, *78*
   pitfalls to avoid, 63–70
   post-initiation phase, 56–57
   project management phases, 53–59
   short- and long-term inputs, 78–79
   short- and long-term outcomes, 79–80
   theory of change, 71–75, *72*
   See also historical considerations for planning
plan views, 125
Plum Village Monastery, 140
policy concerns, 67–68
   See also laws and regulations
political capital
   collaboration among stakeholders and, 163–64
   negative interaction with natural capital, 39–40
   overview, 42–43
pollinators
   bee boxes and beehives, *63*, 138
   habitat in community food forests, 138
   species supporting, 138
polycultures, 27, 111, 114
Ponderosa Park (Basalt, Colorado), 145, 146, 149
popular education, 77
Portland Fruit Tree Project, 21, 213
Powell, Mark, 127
power considerations
   sharing vs. competing, 218
   in systems, 164–65
The Power of Community (film), 193
Powers of Ten (film), 26
P-Patch community garden program, 135, 188–191
production focus, Cultivate Kansas City, 139–141, *139*
professionals, as public space allies
   overview, 121–22
   preparing for conversations, 132
   See also specific fields
Project for Public Spaces, 129
Project Learning Tree, 218

INDEX

project management phases
   closure, 58–59
   establishment, 57–58
   initiation, 54–56
   monitoring and maintenance, 58
   overview, 53–54
   participant turnover, 55
   planning, 56–57
   See also specific phases
property selection. *See* site selection
publicity
   Beacon Food Forest, 18, 181, 183, 191, 192, 193, 220–21
   engaging in dialogue, 220–21
   Zone 4 role in, 179, 181
publicly-owned land, 31–32, 172–73
public policy concerns, 67–68
   See also laws and regulations
public-private partnerships, 31–32
Public Produce (Nordahl), 67
public relations, 179, 181
public spaces
   forms of expression in, 122–23
   landscape as language, 123–24
   need for constructive environment, 121
   role in community food forests, 17–18
   See also allies in public space management
Puget Ridge Edible Park, 184
purpose, of systems, 25, 27–28

Quad City Food Forest, 69, 232
Quell, PJ, 139
questions
   for bridging and bonding, 156
   for collaborative leadership, 203

   initiation phase, 55
   planning phase, 64, 81–82

Rahma Edible Forest Snack Garden, *ci1*, 48, 232
rain events, designing for, 161
raised beds
   hügelkultur, 62, *86*
   Ocean View Growing Grounds case study, *78*, 79
realist role, 175
reassessment, lesson learned, 223–24
recreation, nature as, 89–90
redundancy, benefits of, 27, 29, 31
reflection
   on community, 153
   on historical contexts, 82
   on site selection, 155
   See also community
regional collaboration among community food forests, 122, 226
regulations and laws. *See* laws and regulations
relationships
   importance of, 10
   in social systems, 31–32
   See also social systems; stakeholder development
Remiarz, Tomas, 224
repeating celebrations, power of, 41
researcher role, 175, 176
research on urban food forestry, 225
Research to Action, 74
resilience
   connection to community food forests, 7, 22–23
   redundancy of functions and, 27, 29, 31

resource stocks and flows. *See* stocks and flows of community capitals
reverse mapping, 73
Revive the Roots, 61
Rhode Island, Roger Williams Park Edible Forest Garden, 60–62, *61–63*, 136–37
Rice, Chris, *169*
Richmond Edible Forest Project, 48–49, 165–67, *166*
Richmond Greenway, 165–66
Rifle Correctional Center, 147
riparian buffers, 112, *113*
Roanoke, Virginia, Goodwill community garden, 127–28
Roanoke Community Garden Association, 127–28
Roaring Fork Food Alliance, 145
Rodale, Jerome, 93
Roger Williams Park Edible Forest Garden
   as independent project, 60–62, *61–63*
   wildlife in, 136–37
Roland, Ethan, *33t*, 34
Rural Communities (Flora and Flora), 87

safety concerns
   Dr. George Washington Carver Edible Park, 107–8
   planning considerations, 64–66, *65*
   sector considerations, 116
   urban design contributions, 129
   urban wildlife management, 136–38
Sanctuary for Independent Media, *ci5*, 48, 62–63, 179

INDEX

Schenck, Carl A., 92
Schenk, Daniel, 84, 85
scope and scale of projects
    core group roles, 56, 174–77
    initiation phase considerations, 55–56
    social capital and, 36
    Wetherby Edible Forest case study, 119–120
seasonal celebrations, power of, 41
Seattle, Washington
    about, 184
    P-Patch community garden program, 135, 188–191
    Year of Urban Agriculture effort, 186–87
    See also Beacon Food Forest
Seattle Department of Neighborhoods, 189, 195
Seattle Public Utilities (SPU), 185, 187–88
sectors, in permaculture, 116
seed library (Basalt, Colorado), 144–45, *145*, 149
sense of place, 129
Serenbe Community, 20
Serge, Will, 127–28
servant and authentic leadership, 199–200
1700s and 1800s US food history, 88–91
shaded understory
    Dr. George Washington Carver Edible Park, 106, 107, 221
    establishment phase decisions, 31
    stabilization of food forest composition and, 114
shared leadership, 202
Shenandoah West neighborhood, 127–28

Shiloh community, 101
short- and long-term inputs
    Mercy Edible Park, *133*
    Ocean View Growing Grounds case study, 78–79
short- and long-term outcomes. *See* outcomes
signs
    Basalt Food Park, 146, 147, 149
    Beacon Food Forest, 193
    Hale-Y Community Garden Food Forest, *ci11*
    importance of, *ci8*, *ci9*, *ci10*, 43, 69
    political capital considerations, 42
silvopasture, 112
site assessment
    Goodwill Industries of the Valley community garden, 127–28
    in permaculture design methodology, 124–25
site observation, in permaculture design methodology, 124, 155
site preparation
    benefits of partnerships, 190
    Bloomington Community Orchard, 209
site rehabilitation, 39, 190
site selection
    key questions, 81–82, 156
    natural capital considerations, 39
    overview, 54–55
    perseverance in, 220
    publicly-owned land, 31–32
    reflection on, 155–56
1600s and 1700s US food history, 87–88

6th Ward Garden Park, 173, 177, 234
Slow Food movement, 22
Smart Growth initiatives, 22
Smith, Lincoln, 45
social capital
    bonding and bridging capital, 36, 153–55, *154*, 157
    collaboration among stakeholders and, 163–64
    flow of, 34
    in forming Zone 1 volunteer base, 178
    historical considerations, 85
    as investment, 35–36
    role in collaborative community leadership, 202–4
social design, 7–10
social media
    political capital and, 42
    role in project documentation, 99
    usefulness in organizing people, 181
social patterns, sensitivity to, 156
social systems, 169–181
    adaptive management and, 60
    collaborative community leadership in, 201–4
    community food forests as, 17, 27–28, 31–32
    Fargo Forest Garden case study, 161
    overview, 169–171, *169*
    stakeholder strategies, 171–73, *171*
    Zone 00 - individual, 173–74
    Zone 0 - core group, 174–77
    Zone 1 - consistent participants, 178

INDEX

Zone 2 - occasional participants, 178–79
Zone 3 - peripheral agencies, 179
Zone 4 - one-time engagements, 179, 181
zone template, *180*
societal shifts, 18–20, *19*
soil preparation
  Beacon Food Forest, 190
  Ocean View Growing Grounds case study, 79
Southern Heights Food Forest, 173–74, 233–34
spiritual aspects of community food forests, 140, 141
stacking, in permaculture, 115
stakeholder analysis, 56, 172
stakeholder development
  Basalt Food Park, 148–49
  Beacon Food Forest, 183, 186–191
  Bloomington Community Orchard, 208–12
  capital investments and, 163–64
  community support and, 155, 156, 158–59, 171–73
  Dr. George Washington Carver Edible Park, 106–8
  Fargo Forest Garden, 161–62
  food forest social system, 31–32
  6th Ward Garden Park, 177
  See also leadership; social systems
Stephens-Lee High School, 99–100
Stephens-Lee Recreation Center, 99
stocks and flows of community capitals
  community support and, 158

principles of, 32–34, *33t*
role of collaboration in, 170
storm events, designing for, 161
storyteller role, 175
structural political capital, 42
suburban sprawl, 94
success, lessons learned for, 222–23
Sunrise Rotary Club, 213
sustainability
  agroecology approach, 111–12
  Cultivate Kansas City practices, 140–41
  Dr. George Washington Carver Edible Park practices, 98–99, 102
  food forest benefits, 17–18
  growing momentum of, 21–23
Sustainable Seattle, 193
Swale (floating food forest), 217
synergy, in systems, 26
Syson, Stephanie, 144–45, 146–150
systems map example, *47*
systems thinking, 25–34
  adaptive management and, 59
  characteristics of life-affirming systems, 164, 165
  food forest social system, 31–32
  framework for, 25
  importance of, 24
  need for community support, 158
  nested systems, 26, 29–31, *30*
  power considerations, 164–65
  stocks and flows of community capitals, 32–34, *33t*
  systems map example, *47*
  value of systems concepts, 25–29

target audience, initiation phase considerations, 55
tax incentives, 46
technical skills, 200–201
Teegarden, Aimee, 177
temperate regions
  agroforestry development, 112–14
  food forest layers, 31
Tennessee, Tree Streets Food Forest, 49, 233
terraces and contoured pathways, *ci13*
Texas
  Community First! Village, 20
  Festival Beach Food Forest, 232
  Green Ambassadors program, 218
theory of change, 71–75, *72*, 164
Thich Nhat Hanh, 140
Thoreau, Henry David, 89
Throop Unitarian Universalist Church food forest, *ci3*, 130–31
time manager role, 175, 176
Time-Saver Standards for Landscape Architecture (Dines), 132
Toensmeier, Eric, 160, 238n5
Together Green grant, 139
tragedy of the commons, 192
Transition Pasadena, 130–31
Transition Towns movement, 22, 84
Tree Streets Food Forest, 49, 233
trends in food forests, 224
Tribal Leadership (Logan et al.), 203
tropical home gardens, 112
Troy Bike Repair, 48, 179
turnover in participants, 55

INDEX

United Nations, 224–25
United States School Garden Army, 93
urban agriculturalists, as public space allies, 133–35, *134*
urban agriculture
    Cultivate Kansas City practices, 138–141, *139*
    Seattle Year of Urban Agriculture, 186–87, 188
    See also community gardens
Urban and Community Forestry Assistance Act, 22
Urban and Community Forestry Assistance Program, 132
urban designers, as public space allies, 128–131, *131*
Urban Farm Collective, 161–62
urban food forests, 6–7
    See also community food forests
urban foresters, as public space allies, 132–33
Urban Forestry and Urban Greening (journal), 225
urban gardeners, as public space allies, 135–36
Urban Gardening Program, 21
urban living
    increase in, 132
    mixed progress during the nineteenth century, 91
Urban Tilth, 48–49, 165–67
urban wildlife safety concerns, 146, 172
urban wildlife specialists, as public space allies, 136–38, *137*, 146
US Department of Agriculture (USDA), 21
US food history
    1600s and 1700s, 87–88

1700s and 1800s, 88–91
1900s, 92–94
US Forest Service, 132, 165, 225
US Housing and Urban Development Department, 128

values
    collaborative community leadership and, 202, 203
    cultural capital considerations, 40–41, 147
    natural capital considerations, 40
    systems thinking and, 24
    in Whole Measures Framework, 75
vandalism concerns, 69
Vanderbilt estate, 92, 101
venture capital, 44–45
Victory Garden movement, ix–x, 93
Village Homes (Davis, California), 20
Virginia, Hale-Y Community Garden Food Forest, *ci11*, 41, 136, 179
virtual communities, 181
    See also online presence
visionary role, 175
visioning
    Basalt Food Park, 149–150
    Bloomington Community Orchard, 212–16
    Dr. George Washington Carver Edible Park, 106–7
    examples of, 231–34
    importance in maintaining motivation, 149, 201
    Ocean View Growing Grounds, 77–78

    in permaculture design methodology, 124
volunteers
    empowerment of, 165
    lessons learned, 222–23
    turnover in, 55
    valuing of, 167
    Zone 1, 178
    Zone 2, 178–79
    See also work parties and volunteer days

Walden (Thoreau), 89
Wallace, Caroline, 177
Washington State
    about, 184
    P-Patch community garden program, 135, 188–191
    See also Beacon Food Forest
water sources and flows, mapping of, 116
Watson, Paul, 77, 78
Wetherby Edible Forest, 56, 116, 118–120, *119*, 233
Whitefield, Patrick, 238n5
Whole Measures framework, 74–75, 237n1
Whyte, William H., 129
wildlife habitats, 136–38, *137*
Willie Streeter Community Gardens, 208
windbreaks, 112
work parties and volunteer days
    Basalt Food Park, 149
    Beacon Food Forest, *187*, 194–95, *196*, 222
    Bloomington Community Orchard, 209
    communication about, 67
    human capital considerations, 38

INDEX

lessons learned, 222–23
   Mesa Harmony Garden, *171*
   one-time vs. ongoing, 159
workshops, 28, 38
   See also education

Yamashita, Matt, 146
Year of Urban Agriculture
   (Seattle, Washington), 186–87
Yellowstone National Park, 90
youth involvement
   Basalt Food Park, *ci14*
   Bloomington Community
      Orchard, 214–15
   Dr. George Washington Carver
      Edible Park, 100, 102, 106,
      107, 108
   Green Ambassadors program,
      218
   Richmond Edible Forest
      Project, 49, 165, 166–67
   Sanctuary for Independent
      Media, 48, 63, 179

Zagar, Isaiah, ci5
zones
   application to social systems,
      172–73
   in permaculture, 115–16
   template for, *180*
   Zone 00 - individual, 173–74
   Zone 0 - core group, 174–77
   Zone 1 - consistent
      participants, 178
   Zone 2 - occasional
      participants, 178–79
   Zone 3 - peripheral agencies,
      179
   Zone 4 - one-time
      engagements, 179, 181

About the Authors

Catherine Bukowski is a researcher, communicator, educator, and consultant. She is a PhD candidate at the College of Natural Resources and the Environment at Virginia Tech, where her research focuses on the design and management of community food forests. She has also earned a graduate certificate in Collaborative Community Leadership from the College of Agriculture and Life Sciences at Virginia Tech. Her background is in agroforestry, permaculture, forest ecosystem science, natural resource management, community development, and the creative arts. She is on the board of directors for the Association of Temperate Agroforestry and has participated twice with Partners of the Americas' USAID-funded Farmer-to-Farmer program on agroforestry related projects. Catherine served with Peace Corps in Honduras for three years, where she worked with farmers, women's cooperatives, and the Lancetilla Botanical Garden and Research Center. Her previous research, along with work for an international non-profit, led her to work on forest conservation and agroforestry projects in Brazil, Dominican Republic, Cameron, Columbia, Costa Rica, Ghana, Mexico, and Puerto Rico. For more information about Cathie, visit her website at www.communityfoodforests.com.

John Munsell is an associate professor and forest management extension specialist in the College of Natural Resources and the Environment at Virginia Tech. His background is in sociology and natural resources management. He also is past-president of the Association for Temperate Agroforestry and an associate editor of the journal *Agroforestry Systems*. He has served as a reviewer for the New York City Museum of Natural History, Routledge, Taylor & Francis, and the Social Sciences and Humanities Research Council of Canada. John teaches agroforestry and permaculture at Virginia Tech and has worked with communities from Appalachia to Cameroon to study agroforestry implementation and associated environmental, social, and economic impacts. He has helped design whole-farm plans incorporating permaculture and agroforestry for properties across the United States.

the politics and practice of sustainable living
CHELSEA GREEN PUBLISHING

Chelsea Green Publishing sees books as tools for effecting cultural change and seeks to empower citizens to participate in reclaiming our global commons and become its impassioned stewards. If you enjoyed *The Community Food Forest Handbook*, please consider these other great books related to food production and food systems.

THE FRUIT FORAGER'S COMPANION
Ferments, Desserts, Main Dishes, and More from Your Neighborhood and Beyond
SARA BIR
9781603587167
Paperback • $29.95

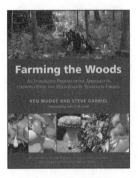

FARMING THE WOODS
An Integrated Permaculture Approach to Growing Food and Medicinals in Temperate Forests
KEN MUDGE and STEVE GABRIEL
9781603585071
Paperback • $39.95

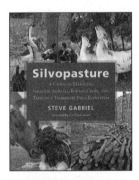

SILVOPASTURE
A Guide to Managing Grazing Animals, Forage Crops, and Trees in a Temperate Farm Ecosystem
STEVE GABRIEL
9781603587310
Paperback • $39.95

FORAGE, HARVEST, FEAST
A Wild-Inspired Cuisine
MARIE VILJOEN
9781603587501
Hardcover • $40.00

For more information or to request a catalog, visit **www.chelseagreen.com** or call toll-free **(800) 639-4099**.

the politics and practice of sustainable living
CHELSEA GREEN PUBLISHING

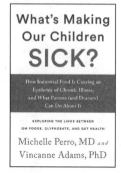

A PRECAUTIONARY TALE
*How One Small Town Banned Pesticides,
Preserved Its Food Heritage, and Inspired a Movement*
PHILIP ACKERMAN-LEIST
9781603587051
Paperback • $19.95

WHAT'S MAKING OUR CHILDREN SICK?
*How Industrial Food Is Causing an Epidemic of Chronic Illness,
and What Parents (and Doctors) Can Do About It*
MICHELLE PERRO and VINCANNE ADAMS
9781603587570
Paperback • $24.95

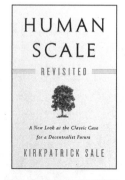

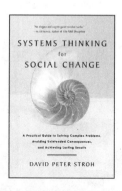

HUMAN SCALE REVISITED
*A New Look at the Classic Case
for a Decentralist Future*
KIRKPATRICK SALE
9781603587129
Paperback • $24.95

SYSTEMS THINKING FOR SOCIAL CHANGE
*A Practical Guide to Solving Complex Problems,
Avoiding Unintended Consequences,
and Achieving Lasting Results*
DAVID PETER STROH
9781603585804
Paperback • $24.95

For more information or to request a catalog,
visit **www.chelseagreen.com** or
call toll-free **(800) 639-4099**.